Arts: A Fourth Level Course

Thought and Reality: Central Themes in Wittgenstein's
Philosophy
Units 7–10

SAYING AND SHOWING
An Introduction to Wittgenstein's *Tractatus Logico-Philosophicus*

Prepared by G. H. R. Parkinson
for the Course Team

The Open University Press

The Open University Press
Walton Hall Milton Keynes

First Published 1976

Designed by the Media Development Group of the Open University.

Printed in Great Britain by

Technical Filmsetters Europe Limited, Manchester, M1 5JY.

ISBN 0 335 05201 0

This text forms part of an Open University course. The complete list of units in the course appears at the end of this text.

For general availability of supporting material referred to in this text please write to the Director of Marketing, The Open University, P.O. Box 81 Walton Hall, Milton Keynes, MK7 6AT.

Further information on Open University courses may be obtained from the Admissions Office, The Open University, P.O. Box 48, Walton Hall, Milton Keynes, MK7 6AB.

1.1

CONTENTS

I   INTRODUCTION

I.1   THE IMPORTANCE OF THE *TRACTATUS*

When Bertrand Russell introduced Wittgenstein's *Tractatus Logico-Philosophicus* to the English-speaking public in 1922, he said of it that it 'deserves, by its breadth and scope and profundity, to be considered an important event in the philosophical world'. The subsequent history of philosophy has proved the soundness of Russell's judgement. The book inspired the so-called 'logical positivists', who flourished in the 20s and 30s and whose ideas made a powerful impact on British philosophy through the medium of A. J. Ayer's *Language, Truth and Logic* (1st ed., 1936). In this book, Ayer declared his debt to Russell and Wittgenstein (op. cit., 2nd, 1946 ed., Penguin Books, 1971, p. 41), and it was the Wittgenstein of the *Tractatus Logico-Philosophicus* of whom he was thinking. But it is not only because of its influence on the logical positivists that the book is important. Wittgenstein came to reject much of what it contained, and such later works of his as the *Philosophical Investigations* can be regarded as a critical commentary on it; indeed, Wittgenstein wanted the *Philosophical Investigations* to be published together with the *Tractatus*, so that the relation between the two works could be made clear (*Philosophical Investigations*, Oxford, 1953, p. *x*). This relation, it may be added, is not one of total contradiction. Although Wittgenstein came to reject much of the *Tractatus*, he did not reject all of it, and some important theses of his later philosophy are to be found in embryonic form in the earlier work.

The importance of the *Tractatus Logico-Philosophicus* is matched by its difficulty. It is a short work of some 70 or 80 pages, but within this short space it deals with wide issues, and (as Russell rightly said) deals with them in a profound way. It is concerned, as Wittgenstein said in the Preface to the book, with the problems of philosophy—not this or that problem, but *the* problems of philosophy—and it claims to provide a definitive solution of these problems. Wittgenstein is able to compress so much into such a small space by presenting his views in the form of a series of concise aphorisms. These are often hard to understand, because they tacitly presuppose a knowledge of a very considerable body of philosophical discussion. The philosophy of the *Tractatus* has been well compared to an iceberg, of which only a relatively small portion is visible; the task of the reader may be compared to that of the sailor, who has to try to relate what he can see to what he cannot, but which he conjectures to be there. This is not an easy task, and it is not surprising that the *Tractatus* has often been misunderstood. But is is not only the conciseness of Wittgenstein's language that is responsible for this misunderstanding. We have spoken of the influence that the book exercised on the logical positivists. These philosophers were in no sense a Wittgensteinian school, mere expositors of another man's ideas; they were original philosophers who took from the *Tractatus* what suited their thinking, and it now appears that in doing so they distorted its doctrines to some extent.

The aim of this correspondence material is to help you to understand the *Tractatus* and, in the process, to clarify your own ideas on some of the subjects with which the work deals. The book is not meant to be a substitute for reading the *Tractatus*, and to this end we shall keep closely to the text. We cannot, in the space available to us, hope to cover everything in the book;

it is hoped, however, that you will be enabled to grasp some of the main philosophical ideas that are contained in it, and to see them in the context of other philosophical thought.

I.2 TRANSLATIONS AND COMMENTARIES

As we are to undertake a close study of parts of the *Tractatus Logico-Philosophicus*, it is important to know something about the English translations. The book was originally published in German in 1921, and entitled *Logisch-philosophische Abhandlung*: literally 'A logico-philosophical treatise' (or 'essay'). The first English translation was made by C. K. Ogden with the assistance of F. P. Ramsey, and published by Routledge in 1922; a revised edition of this translation appeared in 1933. This translation has had its critics, but it has a very important point in its favour: namely, that it was revised by Wittgenstein himself. The changes that he proposed to the proofs of the 1922 edition can now be studied in his *Letters to C. K. Ogden*, ed. by G. H. von Wright (Oxford and London, 1973). The changes (relatively few) made in the 1933 edition were also in response to comments by Wittgenstein. (See C. Lewy, 'A Note on the Text of the *Tractatus*', *Mind*, 1967, pp. 416–423.)

Despite this, the Ogden and Ramsey translation is not wholly satisfactory, and a completely new translation was made by D. F. Pears and B. F. McGuinness, and published by Routledge in 1961. A revised version of this translation was issued by the same publishers in paper-back form in 1974; this takes account of Wittgenstein's comments on the Ogden-Ramsey translation. The revised version of the Pears-McGuinness translation will be used here, and will be referred to by the initial '*T*'. Reference will from time to time be made to the Ogden and Ramsey translation, where this is helpful.

Many books and articles have been written on the *Tractatus*, and although this correspondence material does not follow any one interpreter in particular, it is hardly necessary to say that it owes much to previous work in the field. To comment on commentators, however, would add considerably to the length and complexity of what is intended only as an introduction; consequently, little express reference will be made in what follows to other studies of the *Tractatus*. You may find it useful, however, to have a short list of some general introductions to Wittgenstein's philosophy, to which you can turn for further help. All the books mentioned contain bibliographies.

A. Kenny, *Wittgenstein* (Penguin Books, Harmondsworth, 1975; first published by Allen Lane, The Penguin Press, London, 1973); D. F. Pears, *Wittgenstein* (Fontana/Collins, London, 1971); G. Pitcher, *The Philosophy of Wittgenstein* (Prentice-Hall, Englewood Cliffs, 1964).

I.3 THE HISTORICAL BACKGROUND: FREGE AND RUSSELL

We said earlier that, if the *Tractatus* is to be understood, much that is not stated in the book has to be made explicit. But how are we to do this? How are we to chart the hidden parts of the iceberg? We can derive some help from a number of Wittgenstein's notebooks from the period 1914–16, which contain preliminary work on the *Tractatus* and which, together with some other relevant material, were published in 1961. (Wittgenstein, *Notebooks, 1914–1916*, ed. and trans. by G. E. M. Anscombe, Blackwell, Oxford. This will be referred to henceforth as '*N*'.) But the value of this material is limited. First, it is often as condensed and obscure as the *Tractatus* itself. Second, it consists of *preliminary* work, and (as is often the case with such work) it contains some ideas which were abandoned in the published book. Third, the notebooks do not contain all Wittgenstein's work on the *Tractatus*. Most of the notebooks containing his preliminary work—including a large number from the period when the ideas that culminated in the *Tractatus* were developing—were destroyed by his orders in 1950; the three that have been published survived only because of a lucky accident (*N*, p.*v*). It cannot be assumed, then, that because a certain line of thought is not to be found in the notebooks, it is not present in the *Tractatus*. Nevertheless, the notebooks give some indication of the problems that exercised Wittgenstein in the period immediately preceding the *Tractatus*, and are valuable for that reason.

We can also look for help in the writings of some of Wittgenstein's contemporaries. In his Preface to the *Tractatus*, Wittgenstein acknowledges a debt to Frege and Russell. Like Russell, Gottlob Frege (1848–1925) had been concerned with the foundations of mathematics, which he discussed in a series of important books: *Begriffsschrift* ('Ideography'), 1879; *Die Grundlagen der Arithmetik* ('The Foundations of Arithmetic'), 1884; *Grundgesetze der Arithmetik* ('Fundamental Laws of Arithmetic'), 1893, 1903. He also did important work in the philosophy of logic, particularly in the field of meaning. We shall have occasion to discuss some of this work later; on the whole, however, we shall not emphasize the connexions between the ideas of Wittgenstein and Frege. The books that do so are apt to regard the *Tractatus* as primarily concerned with the foundations of logic (see, e.g., G. E. M. Anscombe, *An Introduction to Wittgenstein's 'Tractatus'*, Hutchinson, London, 1959), and although this is beyond doubt a legitimate approach it is not the easiest one. Our approach will be by way of the philosophy of Russell, and we shall look at the *Tractatus* in the light of Russell's logical atomism. Of particular importance are his lectures on logical atomism, given early in 1918, which will be cited here in the edition by D. F. Pears, *Russell's Logical Atomism*, Fontana/Collins, London, 1972 (abbreviated, *RLA*) and in the edition by R. C. Marsh, *Logic and Knowledge*, Allen and Unwin, London, 1956 (abbreviated, *LK*).

The idea of using Russell's lectures on logical atomism to throw light on the *Tractatus* is obviously attractive. Russell said that in these lectures he was 'very largely concerned with explaining certain ideas which I learnt from my friend and former pupil Ludwig Wittgenstein' (*RLA* 31, *LK* 177), and since the lectures are expansive and lucid whereas the *Tractatus* is condensed and obscure, it is natural to think of using them as a guide to the *Tractatus*. We must be clear, however, about the sort of guidance that they provide. Having worked through Units 5–6 of this course (*Realism and Logical Analysis*), you will already know something of the relations between Russell's lectures on logical atomism and the *Tractatus*, but it may be helpful to give a reminder of the salient points. Russell did not know of the text of the *Tractatus* when he gave his lectures; indeed, the book had not been completed at that time. Russell's personal contact with Wittgenstein had been during the years 1911–13; in 1914 Wittgenstein joined the Austrian army, and Russell lost contact with him until after the completion of the book in August 1918. Strictly, then, the lectures do not comment on the *Tractatus*, but they do relate to a period during which the *Tractatus* was germinating. This raises a further point: namely, the exact relation between Russell's and Wittgenstein's ideas. What was said earlier about the logical positivists and Wittgenstein applies to Russell also. His assertion that he is only explaining certain ideas of Wittgenstein is unduly modest; Russell was a major philosopher in his own right, and he took from Wittgenstein what suited his own thought. The upshot of this is that we cannot apply the text of Russell's lectures to the *Tractatus*, like a key to a cipher, and hope to read off in a more or less mechanical fashion the answers to our questions. We have to use the lectures as a stimulus to further questions: we have to ask, after studying Russell, 'Is this what the *Tractatus* means?' It is perhaps worth adding that this is in many ways the best sort of guide to a philosophical work: a guide which gives help, but which does not remove the necessity for thought.

I.4 OUTLINE

In presenting the *Tractatus*, our usual method will be to begin by specifying a number of passages that are to be read; these will then be discussed, and at suitable points the discussion will be interrupted by exercises (marked off by a bold line) to enable you to master the material that is presented. We will begin with a rough outline of the the way in which the *Tractatus* will be viewed here. At this point you should turn to the text, and read Wittgenstein's Preface.

We have already mentioned, from the Preface, the remark that the book is concerned with 'the problems of philosophy'. The rest of this sentence (p. 3) is of great importance. Wittgenstein says that he believes that these questions have been posed because the logic of our language has been misunderstood. This may suggest that the *Tractatus* is meant to put an end to philosophy, but this is by no means the case. Wittgenstein does maintain, however, that there is a right and a wrong way of philosophizing, and the *Tractatus* is intended to put an end to the wrong way. In effect, he is saying that 'the problems of philosophy' (whatever that term may mean) disappear when we have obtained a proper understanding of the logic of language, but that getting such an understanding is the task of a discipline or activity which may properly be called 'philosophy'. This remained his view, though his opinions about the nature of language were to undergo a considerable change.

We have mentioned the influence exercised by the *Tractatus* on the logical positivists, and indeed what we have just said about the *Tractatus* is strikingly reminiscent of their views. They argued that those philosophical problems which are termed 'metaphysical' are pseudo-problems, generated by a misuse of language, and (as we shall see later) this was Wittgenstein's view in the *Tractatus*. But although there is a similarity between the doctrines of the *Tractatus* and logical positivism, there is also an important difference. This is hinted at in the last paragraph of the Preface, where Wittgenstein says that his book is valuable in two respects. The first (which has already been mentioned) is that the book provides definitive solutions of the problems it considers. The second is that the work shows how little is achieved by the solution of these problems. It is this second point which is important here. Wittgenstein seems to be referring to the account of mysticism and religion given in the concluding pages of the *Tractatus*, an account which differs from that given by the logical positivists. They tended to say that their philosophy put an end to religion, reducing it to a matter of mere feeling. Wittgenstein, on the other hand, says that although the claims of certain philosophers and theologians are false, this does not mean the end of religion. Religion remains, but it has to be seen in a new way.

The difference between Wittgenstein and the positivists can be compared to the difference between Hume and Kant. Both Hume and Kant attacked dogmatic metaphysics; but whereas Hume was sceptical about the claims of religion, Kant directed his attacks, not so much on religion as such, as on certain mistaken ways of trying to defend it. Works such as Ayer's *Language, Truth and Logic* are in the tradition of Hume, and the *Tractatus* has been thought to be in the same tradition. It will be argued later, however (VII.2), that it is closer to Kant.

Broadcasting

There are two radio programmes associated with these units:

Radio 04:   'The Path to the *Tractatus*'. A talk by B. F. McGuinness on the background to some of the central themes of the *Tractatus*.

Radio 05:   'Wittgenstein on "simples"'. A discussion between G. H. R. Parkinson and S. C. Brown of the questions why no examples of 'simples' are given in the *Tractatus*, and what such examples might be.

II WORLD AND FACT

II.1 THE NUMBERING OF THE PROPOSITIONS OF THE *TRACTATUS*

After this rough sketch, we can start our detailed inquiry into the *Tractatus*.
One of the first things that will strike you about the text is the complex
numerical system by which Wittgenstein refers to the various subdivisions of
the work. Decimal numbers are assigned to the individual propositions that
make up the work. The decimal system indicates the relative importance of
the propositions: for example, 2 is more important than 2.1, which is a
comment on 2; 2.1 is more important than 2.11, which is a comment on 2.1;
and so on. Only seven propositions are referred to by whole numbers
without any decimals; that is, there are seven propositions of the first rank.
Each whole number with its attendant decimals (if any) will be referred to in
what follows as a 'number'.

A similar decimal system had already been employed by Russell and
Whitehead in their *Principia Mathematica* (1st ed., 1913), but it would be a
mistake to suppose that the argument of the *Tractatus* is developed in the way
that Russell's and Whitehead's is. *Principia Mathematica* is a deductive system,
in which theorems are derived from axioms ('primitive propositions') by
means of rules of derivation; there is nothing of this about the *Tractatus*.
Further, each axiom and theorem of *Principia Mathematica* is regarded as
understandable in itself, once we have grasped the technical notation.
Obviously, to prove a theorem we have to see it in the light of the axioms
and theorems which precede it, but we can understand it without seeing it in
this way. In the *Tractatus*, on the other hand, some technical terms are
introduced without a full immediate explanation; to grasp their meaning, we
have to see them in the light of propositions which follow. (See, for example,
the reference to 'logical space' in *T* 1.13, which is not explained until 2.11.)

It might be supposed that we could divide our discussion of the *Tractatus* in
accordance with the numbers of the work, treating each number as a chapter
and assigning a section of the discussion to each. In fact, however, the logical
subdivision of the argument of the book does not correspond exactly to the
various numbers. This can be seen almost at once. The *Tractatus* begins with
a series of propositions about facts, states of affairs and objects. These occupy
the whole of number 1 and a part of number 2, ending at 2.063. As these
topics are closely inter-related, and as they are followed in 2.1 by what is
clearly a new topic—namely, pictures of facts—they will be treated here as a
unity and will form the subject-matter of this section.

Many of the assertions to be discussed in this section may seem to be merely
dogmatic. In fact, this is not so; Wittgenstein has his reasons for what he
says, but many of these are not presented until later. However, the
presentation of these reasons presupposes the technical vocabulary that is
employed early in the *Tractatus*, and so it will be most convenient to follow
Wittgenstein's order of exposition, and to defer for the moment his reasons
for what he says.

II.2 FACTS

Reading: *T*1–2.01

It will be seen that the first nine propositions of the *Tractatus* deal with 'facts'. The seven propositions that make up number 1 state that the world is everything that is the case; that it is the totality of facts, not of things; and that it divides into facts. The first two propositions of number 2 say that a fact is the existence of states of affairs, and that a state of affairs is 'a combination of objects (things)'. All this is the tip of an iceberg; what is underneath it? What exactly does Wittgenstein mean here? With what problems is he dealing? Let us see what guidance we can derive from Russell's *The Philosophy of Logical Atomism*.

Before going further, read what is said about facts, and the complexity of a fact, in *RLA* 35–7 and 47 (*LK* 182 f., 192 f.).

In his opening lecture, 'Facts and Propositions', Russell says that a fact is what makes a proposition true or false (*RLA* 36 f., *LK* 182 f.). At this stage, we will not try to give a full account of what is meant by 'proposition'; we will simply take it that a proposition has the characteristic of being true or false, that it is (*RLA* 38 f., *LK* 184 f.) the vehicle of truth and falsehood. There will be much more to say about propositions in later sections; for the moment, our concern is with facts. Facts, Russell insists, belong to the objective world. They 'are what they are whatever we may choose to think about them' (*RLA* 35, *LK* 182). A fact, Russell adds, is not a thing; Socrates, for example, does not make the proposition 'Socrates is dead' true. What makes the proposition true is *the fact that* Socrates is dead.

Russell clearly has in mind here something resembling the correspondence theory of truth. It so happens that in *The Philosophy of Logical Atomism* Russell does not state this theory in its usual form, according to which a proposition is true if it corresponds with a certain fact, and false if it does not. Instead, he says (*RLA* 64, *LK* 208) that a proposition can correspond with a fact in either a true or a false way. But in earlier works (to which, incidentally, Wittgenstein could have had access before he wrote the *Tractatus*) Russell said that correspondence with fact constitutes the nature of truth (*The Problems of Philosophy* (1912), Oxford Paperback Edition, p. 71; cf. 'On the Nature of Truth and Falsehood' (1910), in *Philosophical Essays*, 1966 ed., p. 158).

Let us take it as a hypothesis that Wittgenstein had something similar in mind in the first nine propositions of the *Tractatus*. If this hypothesis is correct, they can be read as follows. What make propositions true or false are facts, not things; to speak of truth is to speak of a correspondence of proposition with fact. In calling the totality of facts 'the world', Wittgenstein — *correspondence* is emphasizing the point that facts are objective, i.e. not mind-dependent.

Our hypothesis can be supported by passages from other parts of the *Tractatus*. Wittgenstein does not say in so many words that a true proposition is one which corresponds with the facts, but he does say (4.01) that a proposition is a picture of reality, that (2.21) a picture either agrees or disagrees with reality, and that (2.222) truth consists in the agreement of its 'sense' with reality. These propositions, it will be noticed, speak of an agreement with 'reality', but Wittgenstein says (2.063) that the world is the sum total of reality, thus linking these propositions with his earlier assertion that the world is the totality of facts.

In sum, it appears that the opening propositions of the *Tractatus* have to be seen in the context of the rebellion against idealism, and in favour of realism, led by Moore and Russell at about the turn of the century. In these propositions, Wittgenstein is ranging himself on the side of the realists. One word of caution: the fact that we refer at this early stage to the nature of truth does not mean that the problem of truth is the central issue of the *Tractatus*. On the contrary, Wittgenstein's chief concern in this book is with meaning rather than with truth. Still, he is concerned with truth and falsity, and it will be seen later that when he discusses meaning in the *Tractatus*, he constantly has in mind the problem, 'How are *false* propositions possible?'

So far, we appear to be on the right track; but we have further to go. Let us consider again *The Philosophy of Logical Atomism*. We have seen that Russell says that a fact is what makes a proposition true or false; he also says (*RLA* 47, *LK* 192 f.) that all facts are complex, each one having certain components. There are, he says (*RLA* 53 f. *LK* 199), 'facts in which you have a thing and a quality, two things and a relation, three things and a relation, four things and a relation, and so on'. When Russell says, then, that 'correspondence with fact' constitutes the nature of truth (*The Problems of Philosophy*, p. 71), he means that truth consists in a correspondence with something complex; indeed, he sometimes speaks of a correspondence with *a* complex, saying (op. cit., p. 74) that a belief is true when it corresponds to a certain associated complex.

We now have to see whether it is also the doctrine of the *Tractatus* that facts are complex. Wittgenstein does not say explicitly that a fact is complex, having certain components, but he does say that a fact is 'the existence of states of affairs' (*T* 2), and that a state of affairs (*Sachverhalt*) is a combination of what he calls alternatively objects (*Gegenstände*) or things (*Dinge*) (2.01). A state of affairs, then, is clearly complex; but we wanted to know whether Wittgenstein thought that *facts* are complex. The short answer is that it is hard to see how facts can be other than complex; for facts are what is the case, and what is the case is the existence of states of affairs, and states of affairs are complex. This answer, however, may leave one dissatisfied; one still wants to know just *how* facts are related to states of affairs.

Russell put a similar question to Wittgenstein, who answered it in a letter (19.8.19; *N*, p. 129). The reply will not be wholly clear at present, since it involves a technical term, 'elementary proposition', which we have not met so far; however, the general sense will be clear enough. The difference between a fact and a state of affairs, Wittgenstein says, is that a state of affairs is what corresponds to an elementary proposition if that proposition is true; a fact is what corresponds to the logical product of elementary propositions when that logical product is true. It will be noticed that this reply confirms our suggestion that Wittgenstein's views about facts have to be seen in the light of a theory of truth; what concerns us at present, is what it says about the relation between facts and states of affairs. We may, for the moment, leave aside the question, 'What is an elementary proposition?' (this will be considered later, in VI.1); it is sufficient for us to regard an elementary proposition as a proposition of an unspecified kind. We must, however, see what is meant by a 'logical product' of propositions. In *Principia Mathematica* (Introduction, 2nd ed., Cambridge 1927, p. 6) the 'logical product' of two propositions, p and q, is said to be the proposition that asserts that both p and q are true; similarly, the logical product of propositions p, q and r will be the proposition that asserts that p, q and r are all true; and so on. (Students who have followed the Foundation Course, A 100, *Introduction to Logic*, will have met this notion—though not under this name—in those pages of Part VII that deal with translating '&'.) From this it follows that, if $S1$ is the state of affairs corresponding to the true elementary proposition $p1$, and $S2$ is the state of affairs corresponding to the true elementary proposition $p2$, and $S3$ is the state of affairs corresponding to the true elementary proposition $p3$, then what corresponds to the logical product of $p1$, $p2$ and $p3$ is what Wittgenstein calls a 'fact'. In short, a fact is a collection or set of states of affairs. It is clear from this that a fact is complex, and how it is complex; it is complex in that it is a collection or set of states of affairs, each of which is a combination of objects.

Since a fact is a set of states of affairs, we may expect to be told more about these states of affairs, and this is what Wittgenstein does in a series of propositions (2.011–2.063) which comment on the assertion (2.01) that a state of affairs is a combination of objects. This part of the *Tractatus* will be our next concern.

II.3   OBJECTS AND STATES OF AFFAIRS

Reading: T 2.011–2.02, 2.03–2.063

We have seen that a fact is a set of states of affairs, and that (2.01) a state of affairs is a combination of objects. It follows that facts, too, are combinations of objects. The way in which a fact is to be distinguished from a state of affairs may be explained as follows. A fact, as we have seen, is a combination of objects, such that it can be subdivided into other combinations of objects. If these sub-groups are such that none of them can be subdivided into other combinations of objects, then each such sub-group is a state of affairs. In a sense, then, a state of affairs is atomic, in that it is a combination of objects such that it cannot be subdivided into other combinations of objects. This point is brought out by the way in which the term *Sachverhalt* is rendered in the older translation of the *Tractatus*. This term, translated by Pears and McGuinness as 'state of affairs', was rendered by Ogden and Ramsey as 'atomic fact'. This is not a literal translation of the German, but it is worth noting that it was sanctioned by Wittgenstein himself (*Letters to C. K. Ogden*, p. 59).

The term 'atomic fact' is reminiscent of Russell's logical atomism, and since the propositions in which Wittgenstein speaks of 'states of affairs' are often far from clear, it may be helpful to approach them by way of Russell. The likenesses, and also the differences, between the two views may help to bring out more clearly what Wittgenstein means.

Let us first remind ourselves of the salient features of Russell's views about atomic facts. Such facts, he says, are of various kinds (*RLA* 53 f., *LK* 198 f.). Some consist in the possession of a quality by a particular thing; some in a 'dyadic' relation between two things; some, in a 'triadic' relation between three things; and so on. For the sake of convenience, Russell proposes to call a quality a 'monadic' relation. This enables him to say that every atomic fact contains a relation and the term or terms of the relation—one term if the relation is monadic, two if it is dyadic, and so on. But although this distinguishes between the various types of atomic facts, it does not explain what an atomic fact *is*. Our position so far is like that of someone who has been told that some games involve one individual, some involve two, some involve three, and so on, but who knows no more about games than that. Russell makes his meaning clearer after he has said that the terms which enter atomic facts may be called 'particulars'. In everyday life, he says, we accept as particulars what are not really such. Socrates, for example, is not a particular; or, as Russell prefers to say, the word 'Socrates' does not stand for a particular. Really, the word is an abbreviation for a description; a description, not of a particular, but of a system of classes (*RLA* 55 f., *LK* 200 f.). It is clear from this that if a word stands for a class of any kind, then what that word stands for is not a particular. In other words, the particulars which are the terms of atomic facts are simple, not composite.

With this in mind, let us approach what Wittgenstein says. We shall refer to a *Sachverhalt* as a 'state of affairs', since this is the term used in the translation that we are studying, but it should be borne in mind that 'atomic fact' is an alternative translation. Now, we have spoken of Russell's atomic facts; to what extent do Wittgenstein's 'states of affairs' correspond to these?

Exercise

Before going further, re-read the passages set for this sub-section, and Lecture II of Russell's *The Philosophy of Logical Atomism* (*RLA* 44–57, *LK* 189–202). Then give your answer to the question, 'What are the likenesses, and the differences, between the views stated in these passages from the *Tractatus* and Russell's logical atomism?'

There is one striking similarity. We said just now that Russell's particulars could be called simple; in *T* 2.02 Wittgenstein states this expressly of his objects. However, we are not yet in a position to answer the question whether Wittgenstein would have agreed with Russell that Socrates is not something simple. It may seem obvious that Socrates is not something simple; but some philosophers—Leibniz, for example—would have said that this is not so, in that Socrates is really a soul, and souls are simple. Later (IV.5.1), after an examination of Wittgenstein's views about meaning, we shall be able to discuss this question, and we shall see that Wittgenstein would have agreed with Russell on this point.

But there are also some important differences between Russell's atomic facts and Wittgenstein's states of affairs.

(i) We said that Russell's atomic facts contain particulars; but Russell also says that each atomic fact contains a relation as well (*RLA* 54, *LK* 199; see also *RLA* 33, *LK* 179). Wittgenstein, on the other hand, says that in states of affairs, objects fit into each other 'like the links of a chain' (2.03). The meaning of this remark, as Wittgenstein explained to his translator (*Letters to C. K. Ogden*, p. 23), is that objects are not connected by anything other than themselves; rather, the objects themselves make the connexion. This could be read as a comment on Russell. In *The Problems of Philosophy*, Russell had said that relations can act as a kind of cement, binding objects together. For example, if it is true that Desdemona loves Cassio, then there is a complex unity—a fact—in which the relation of loving binds together the objects Desdemona and Cassio (Russell, op. cit., p. 73). What Wittgenstein has said so far does not commit him to saying that a relation is not an object (we shall see his views on this later, in VI.1.3). But he would say that if it is an object, there is no need for that special kind of object called a relation to join other objects in states of affairs.

(ii) A second difference is equally important, though perhaps less obvious, and concerns problems of thinking and knowing. The relevant passages in the *Tractatus* (2.011–2.0141) deal with the relations between things (objects) and states of affairs. We know already that states of affairs are combinations of objects or things. Now, Wittgenstein says that it does not just *happen* that a thing can be a constituent of a state of affairs. Rather, this is essential to a thing (2.011); a thing *must* be capable of being a constituent of a state of affairs (2.0121). These remarks seem easy enough to understand, but it is less easy to see their point. As often, the *Tractatus* gives us an answer without first stating a question. The question, hinted at in 2.0121, is: 'What is it to think of an object? More specifically, can one think of an object by itself, abstracting from its capacity to combine with other objects in states of affairs?' Wittgenstein's answer is that one cannot. Similarly (cf. 2.0123) one cannot know an object without knowing all its possible occurrences in states of affairs.

Let us try to give an example of what Wittgenstein means. For illustrative purposes, let us suppose that musical notes are objects in Wittgenstein's sense, and let us ask: 'What is it to think of a particular musical note—say, middle C?' One might be tempted to answer that the person who is thinking of middle C can imagine that note, and imagine *just* the note and no more. Wittgenstein would reply that more is required. The person in question is supposed to be thinking of *middle C*, and this means that he must be able to think of middle C as occurring in states of affairs: for example, the states of affairs to which there correspond the true propositions 'Middle C is a musical note', 'Middle C is a whole tone below the D immediately above it', and so on. This differs from Russell's views in *The Philosophy of Logical Atomism*. Russell did not expressly consider the case of someone thinking of a particular, but he did consider what is involved when the word for a particular has meaning for someone. Russell declared (*RLA* 57, *LK* 202) that in order to understand the name for a particular, the only thing necessary is to be acquainted with that particular. From this it seems to follow that (if middle C is a particular) the person who understands the term 'middle C' can think of middle C in isolation from everything else: as Russell put it, 'No further information is required'. It is clear that Wittgenstein would say that further information is required, and we shall find that this is confirmed when we discuss later his views about names (IV.6 below).

(iii) According to Wittgenstein, states of affairs are independent of each other; from the existence or non-existence of one state of affairs we cannot infer the existence or non-existence of another (2.061, 2.062). The importance of this thesis can perhaps best be shown by an example. Suppose that one state of affairs (or, it may be, one set of states of affairs) is the application of a flame to paper, and suppose that another state of affairs (or set of states of affairs) is the ignition of that paper. One would be inclined to say that the two are causally connected, and that it is possible to infer the one (the effect) from the other (the cause). Wittgenstein would deny that it is possible to infer in this way; there is, he says, no 'causal nexus' (5.136; cf. VI.4 below).

Russell does not say expressly that atomic facts are independent of each other, and indeed it seems that he held the opposite view. This has to be inferred from what he says about atomic propositions, which implies that these are not independent of one another (D. F. Pears, *Bertrand Russell and the British Tradition in Philosophy*, London, 1967, p. 141). Now, since a true atomic proposition corresponds to an atomic fact it follows that, for Russell, atomic facts are not independent of each other.

II.4   OBJECTS AND SUBSTANCE

Reading: *T* 2.0231–2.0272

There remains for discussion another important view about the nature of objects that is put forward in the *Tractatus*, and this can again be introduced by way of a comparison with Russell. A particular, according to Russell, 'has the sort of self-subsistence that used to belong to substance, except that it usually only persists through a very short time, so far as our experience goes' (*RLA* 57, *LK* 202). Russell is here making two points. He is saying that each particular is self-subsistent, by which he means (loc. cit.) that it does not depend logically on any other; he is also saying that each particular persists, i.e. it endures or lasts.

The term 'substance' first appears in the *Tractatus* in 2.021, in which Wittgenstein says that objects make up the substance of the world. He gives no hint about the way in which he is using the word 'substance', and it may be that this particular proposition is not meant to add to our understanding of the concept of an object, but to remind us of what we have already learned. It will be remembered that for Wittgenstein, the world is everything that is the case; it is the totality of facts. A fact, in turn, is the existence of states of affairs, and a state of affairs is a combination of objects. To say that objects make up the substance of the world is simply to remind one of this; it is to remind us that *what* are combined to form states of affairs are objects. Substance, one might say, is what there is, as opposed to how things are.

Russell said of particulars that, like substances, they persist. Similarly, Wittgenstein says in 2.024 that substance is what subsists independently of what is the case. Since objects make up the substance of the world, 2.024 should be a way of saying that objects subsist; and indeed Wittgenstein remarks (2.0271) that objects remain the same—what changes is their 'configuration', that is, the way in which they are combined.

Exercise

It was said in the last sub-section that objects must be able to occur in states of affairs (cf. 2.011). Does this contradict the assertion (2.024) that substance —that is, objects—is what subsists independently of what is the case?

There is in fact no contradiction here. Objects must indeed be able to occur in states of affairs, but there is no state of affairs of which any object must remain a member. One may make a rough comparison with a child's construction set, which consists of a number of pieces of metal, each of which may be joined with others in various ways. Each piece can occur in states of affairs (i.e. can be joined with others in various ways), but the pieces remain the same when the combinations of the pieces change.

The other feature of Russell's particulars mentioned earlier was that they are self-subsistent, in that no one of them depends logically on any other. Russell seems to mean by this that it is not possible to infer from one particular to another, and he takes this to imply that one cannot say that because there is one particular, therefore there is any other particular. It is, in his view, logically possible that there should be just one particular in the universe, though as a matter of empirical fact this is not so (*RLA* 57, *LK* 202). Let us consider what the *Tractatus* would say about this last point. We have seen, in (ii) on p. 14, that objects must be capable of occurring in states of affairs; that is, they must be capable of combining with other objects. But this does not, of itself, imply that they do. Could there, according to Wittgenstein, be just one object?

To see Wittgenstein's answer, we must turn back to 2.0122. In this proposition, he asks in what respect an object can be 'independent'. (The term that he uses is *selbständig*, a word that can also be rendered as 'self-sufficient', 'self-dependent', and is clearly close in meaning to Russell's 'self-subsistent'.) He answers that in one respect an object is independent, but that in another it is not. It is independent, he says, in so far as it 'can occur in all *possible* situations'. He does not define the term 'situation' here, but it emerges from what is said later (cf. III.2) that the term is roughly equivalent to 'fact'. An object, then, is independent in so far as it can occur in all *possible* facts. By stressing the word 'possible', Wittgenstein indicates that an object is not bound to any *actual* fact; that is, it does not have to occur as a member of any one specific fact. However (Wittgenstein goes on), this form of independence is also a form of dependence, in that it is a form of connexion with states of affairs. This appears to mean that, though an object is not bound to any one fact, it must occur as a member of some fact or some state of affairs. As a rough analogy, we might compare a certain sort of dance, in which the dancers constantly change partners, but each dancer must always have a partner. Each dancer may be called independent, in that he or she is not restricted to any one partner throughout the dance; on the other hand, no dancer is independent, in so far as a dancer is never allowed to dance a solo, but must always dance with a partner. We said that the analogy is only rough: it breaks down in that the rules of the dance are purely conventional, whereas in speaking of objects and their independence, Wittgenstein is speaking of what *must be* the case.

It is clear from this that Wittgenstein would not agree with what Russell says about the 'self-subsistence' of particulars; he would not say that it is logically possible that there should exist just one object. An object, for Wittgenstein, must be a constituent of *some* state of affairs; that is, it must be one of a combination of objects. We saw in the last sub-section that an object must be able to combine with others; we have now seen that there must be other objects with which it can combine. In this respect, it can be misleading to compare Wittgenstein's objects with the pieces of a child's construction set, as we did earlier. Consider, for example, an inventor making a prototype model of such a set. After he has made the first piece, it makes sense to say that this piece can be joined with others after they have been made, but that at present it is the only piece. We have seen, however, that it is wrong to suppose that there could be just one Wittgensteinian object, which could combine with others if they existed, but which is in fact the only object.

II.5   OBJECTS AND FORMS

Reading: *T* 2.0123–2.01231, 2.0141, 2.0231–2.0233

There remains for consideration at present one further feature of objects. Wittgenstein says in 2.0231 that the substance of the world determines, and indeed can only determine, a form, and not any material properties. We know already (2.021; cf. II.4 above) that objects make up the substance of the world, so 2.0231 is stating another feature of objects: namely that they determine, and can only determine, a form. Since Wittgenstein speaks elsewhere (*T* 2.0141) of the form of *an* object, he may be taken to mean that *each* object determines a form. That is, the determination of a form is not like a declaration of war (which is done by a country as a whole, and not by each citizen individually); it is more like the consumption of food, which is done by each member of a country.

But what is meant by saying that an object 'determines' a form, and what is a form? In normal usage, *A* may be said to determine *B* if *B* follows necessarily from *A*, whether by virtue of the laws of logic or of science. So, for example, a circle's radius is said to determine its area; that is, given that its radius is such and such, its area *must be* such and such. In the present passage of the

Tractatus, Wittgenstein is saying that if X is an object then it must have a form, but that there is no necessity for it to have this or that material property. What is meant by 'form' here is explained by 2.0141: the form of an object is the possibility of its occurring in states of affairs. Wittgenstein has said earlier (2.0123; cf. II.3 above) that if I know an object I also know all its possible occurrences in states of affairs; he is now saying that this is *all* that I know of X if I simply know that X is an object. The precise nature of the 'material properties' of an object must be deferred for the present (the question is taken up again in IV.5). It will already be clear, however, that they have to do, not with mere possibility, but with what is actually the case.

What has just been said about an object's determination of a form may seem to be a specialized point, of no general philosophical interest. In fact, however, what Wittgenstein is saying here concerns a philosophical point of great importance, which relates to what are often called 'logical types' or 'categories'. You may be familiar with the central position that the notion of a category occupies in Ryle's philosophy of mind; you may know how Ryle argues that the Cartesian view about the relations between mind and body rests on 'category mistakes'. Russell had previously drawn attention to similar errors, which rested on a confusion of what he called 'logical types'. (*RLA* 112 ff., *LK* 254 ff. is one of Russell's many expositions of his theory of types.) In the part of the *Tractatus* that we have just been considering, Wittgenstein has been explaining what it is for an object to be of a certain logical type or category, or, as he says, to be of a certain 'logical form' (2.0233). Roughly, his view is that an object's being of a certain logical type is its ability to enter into such and such combinations of objects.

It is important to avoid a possible misunderstanding here. What has just been said might be thought to imply that Wittgenstein accepted Russell's theory of types; but in fact he did not (cf. T 3.331). This may well seem puzzling. How, it may be asked, can Wittgenstein have a use for the notion of a logical type, and yet reject Russell's theory of types? For reasons that will become clear shortly, we are not yet in a position to give a full answer to this; however, what can be said at present is this. Russell's theory of types was meant to provide a way of escape from certain logical paradoxes, of which we may mention one that concerns classes. Russell pointed out (*RLA* 119 f., *LK* 260 f.) that it might seem meaningful to ask whether or not a class is a member of itself. It might also seem meaningful to say that 'in all the cases of the ordinary classes of everyday life … a class is not a member of itself'; for example, the class of men is not a man. But (Russell went on) consider the class of all classes that are not members of themselves. Is this a member of itself, or not? If it is a member of itself, it is not, and if it is not, it is. Russell's solution was that 'a totality of any sort cannot be a member of itself' (*RLA* 123, *LK* 264). A class and the individuals that are its members are of different types, and what can be said meaningfully of the one cannot be said meaningfully of the other. It is meaningful to say, for example, that Socrates is a man, but to say that the class of men is a man is to confuse logical types, and is meaningless. Similarly, a class and a class of classes are of different types; so, too, are a class of classes and a class of classes of classes— and so on for an infinite hierarchy. The paradox about the class of all classes that are not members of themselves rests on a confusion of types, and so is meaningless.

Wittgenstein's objection to the theory of types was that it set out to do what cannot be done: namely, to state the logical structure of meaningful discourse. Russell, as Wittgenstein would put it, was trying to say what logical form is; but this is something that can only be shown. Wittgenstein

would agree that there are in a sense differences of type, but he would argue that such differences can only be shown; we cannot *say* what they are. All this cannot but seem obscure. To mention logical form, and the distinction between saying and showing, is to anticipate matters that can be discussed in detail only later (see III.2, III.3 and especially V.4). However, it was necessary to make some reference to them here so as to avoid a possible misunderstanding of Wittgenstein's views about the form of an object.

II.5.1 What Would be Examples of Objects?

We have now seen various features of objects: they must subsist independently of what is the case, they must be able to be constituents of states of affairs, they must determine only a form and not any material properties. It will be noticed that Wittgenstein says, not that they just happen to have these features, but that they must have them. This has a bearing on a question which it is natural to ask, namely '*What* has all these features? What would be an example of an object, as that term is understood in the *Tractatus*?' Wittgenstein does not answer this question. There is no reason to believe that he thought that no examples of objects could be given; he simply thought that it is not a philosopher's business to settle the question whether a given thing is simple or not (N. Malcolm, *Ludwig Wittgenstein: A Memoir*, Oxford, 1958, p. 86). That question, he said, is an empirical one; the philosopher's inquiries, on the other hand, are not empirical. (This point will be taken up in VII.3 below.) The philosopher's business is to settle, independently of experience, the criteria that an object must satisfy; it is not his business to say what actually satisfies these criteria. Russell, incidentally, put forward the same view in *The Philosophy of Logical Atomism*, arguing (*RLA* 54, *LK* 199) that the question of what particulars are actually found in the real world is a purely empirical one, which does not interest the logician as such. Nevertheless, the question 'What would count as examples of objects, as these are described in the *Tractatus*?' is one which it is hard to ignore. At present, however, we have not seen enough of the *Tractatus* to be able to discuss this question; in particular, we have not yet examined Wittgenstein's views about meaning, which are of fundamental importance here. We shall return to the question much later, in VI.1.3.

So far, we have been trying to see what is *meant* by the propositions about objects between *T* 1 and 2.063; we have said nothing about *why* these propositions are asserted. We have not asked:

1 Why Wittgenstein thinks that there *are* objects of the kind that he describes.

2 Why it is said that objects cannot determine material properties.

Wittgenstein answers these questions early in the *Tractatus*; question 1 is answered in 2.0201–2.023, and question 2 in 2.0231. However, neither answer is fully intelligible in the light of what we have learnt so far, in that each (like the answer to the question, what would be examples of objects) involves a theory of meaning. It is now time to discuss this theory, and to move from Wittgenstein's account of facts to his account of the nature of the proposition.

III PICTURES

III.1   PICTURES: THEIR ELEMENTS AND STRUCTURE

Reading: *T* 2.1–2.1515

After a promise of a discussion of the *Tractatus*' views on propositions, a discussion of pictures may come as a surprise. In fact, however, we are following the order of the *Tractatus*, and it is a logical order. The account of pictures that we are to discuss, which begins at 2.1 and covers the rest of number 2, is the first stage of a long and complex discussion which continues with an account of the nature of the proposition in number 3. In number 4, Wittgenstein puts together the results of these previous inquiries, and speaks about propositions as pictures.

Wittgenstein begins his account of pictures by saying (2.1) that we picture facts to ourselves. This needs a good deal of comment.

(a) We are said to *picture* facts to ourselves. The translation may suggest that Wittgenstein is referring to mental images, pictures seen by 'the mind's eye'. We shall see later that there is some reason to suppose that he would not refuse to count mental images as pictures. (To Russell, 19.8.19; *N*, p. 130. Cf. IV.1.) On the other hand, it is quite certain that not all pictures, as Wittgenstein understands the term, are mental images: this is shown, for example, by the reference to hieroglyphics and alphabetic script in 4.016. In this respect, then, the more literal translation of Ogden and Ramsey is preferable: 'We make to ourselves pictures of facts'.

Now, leaving aside mental images for the moment, what are these pictures? The reference to hieroglyphics (i.e. picture-writing of a kind) that we have just cited suggests that at least some of the pictures are pictures in the ordinary sense of the word; that is, they are drawn or painted on a flat surface. On the other hand, the reference to a *tableau vivant* in 4.0311 suggests that a picture may be three-dimensional, and in the Notebooks Wittgenstein referred to the way in which a motor-car accident is represented by means of dolls etc. (*N*, p. 7). Further, in 2.182 Wittgenstein says that not every picture is even a spatial one.

There seem, then, to be many different examples of pictures, as Wittgenstein understands the term, and it is not immediately obvious what makes them pictures. However, we have seen that Wittgenstein's pictures include what would normally be regarded as pictures, and it will simplify our account if we concentrate on these for the time being; we can widen the concept of a picture later, once we have seen something of Wittgenstein's meaning in the special case of what is painted or drawn.

(b) We are said to picture, or make pictures of, *facts*. It will be remembered that 'fact' has a technical sense in the *Tractatus*, and we are quickly reminded of this when Wittgenstein comments on 2.1 by saying (2.11) that a picture presents 'the existence and non-existence of states of affairs'. However, nearly all the discussion of pictures in number 2 is at the level of facts, not of states of affairs. States of affairs, then, and the objects which are their constituents, can be left aside for the moment; they will meet us again when we consider the proposition in section IV. What is important to remember about facts here is that a fact is what makes a proposition true or false (cf. II.2). In saying, then, that we make pictures of facts, Wittgenstein implies that such pictures can be called true or false; and indeed this is expressly stated in 2.21. One might put the point as follows: to picture a fact is, in a way, to *say* something—and not in the sense in which an artist may say something in that he expresses an *attitude* by his picture, but in the sense of saying something true or false.

This means that not everything that would normally be called a picture is a picture in Wittgenstein's sense. It is not difficult to see that many pictures ('pictures' in the ordinary sense) are such that it would not be appropriate to say of them, 'Yes, that's true; that is how things are (or were)', or again 'No, that's not true'. Suppose, for example, one looks at a coaching scene on a Christmas card; it would hardly be appropriate to say, 'But this never happened; just *these* people at an inn never greeted just *these* visitors'. If one did, the reply would be 'Whoever said or implied that this did happen? This is just a picture of a coaching scene.' On the other hand, there are pictures in the ordinary sense of the word of which it would be appropriate to say that they are true or false; pictures of famous historical events, say. It is this kind of picture with which we are concerned here.

Exercise

Students of Course A303, Units 14–15 (*Philosophy of Language I*, pp. 12–13) will recollect that in the *Philosophische Grammatik*, p. 164, Wittgenstein contrasts two sentences about pictures: 'This picture represents people in a village inn' and 'This picture represents the coronation of Napoleon'. He says that a picture of the former type 'says something to him', even though he does not believe that the people in it are real people, or that there ever were real people in this situation. Of the two contrasted pictures. which is a picture in the sense understood in the *Tractatus*, and why?

We said that a picture (as 'picture' is understood in the *Tractatus*) can be true or false. For this reason, the picture of people in a village inn is not a picture, in the *Tractatus*' sense of the word. But let us think for a moment about false pictures. Suppose, for example, that it is a fact that a certain cat is sitting by the side of a certain mat. Suppose, further, that someone pictures this fact incorrectly; e.g. he produces a picture which it would be normal to call a picture of the cat (that particular cat; not just any cat) on the mat (that particular mat; not just any mat). Now, to say that he has produced a picture of the cat on the mat is perfectly legitimate in many contexts, but in the context of the *Tractatus* it could cause confusion. Wittgenstein has said that we make pictures of facts; but in the present case, the picture is not a picture of the fact that the cat is on the mat, because the cat is not on the mat. To avoid confusion, then, let us say that the person in question has produced a picture which *means that* the cat is on the mat. This raises an important question: how does this picture mean that the cat is on the mat? Much of the discussion of pictures that follows in number 2 of the *Tractatus* can be seen as an answer to this.

Now that we have roughly identified what is understood in the *Tractatus* by a picture, let us go further into the notion. For the time being, let us continue to concentrate on those pictures which are also pictures in the ordinary sense—that is, drawings and paintings. Suppose, then, that someone pictures the fact that a certain cat is on the mat, and does so by drawing a picture which means that the cat is on the mat. This picture is, in this case, a true one, but its truth is not what concerns us here. We want to know, what makes it a picture, in the sense in which this is understood in the *Tractatus*?

The first thing to note is that there is no one sketch that the man need produce. He may be an expert artist, and produce a lifelike sketch; he may be quite unskilled in drawing, and produce a crude scrawl. Yet both may be equally suitable, provided that certain conditions are satisfied. What, then, are these conditions?

We know already that a fact (like a state of affairs) is a combination of objects. For the sake of illustration, let us suppose that the cat and the mat can both be counted as objects. In fact, it is pretty certain that they are not objects, in Wittgenstein's sense of that term. However, we are (as we said) not aiming at complete precision at present; our aim is to begin by getting a rough idea of what is meant, and to refine on this later (see section IV). Our assumption will do no harm, provided that we remember that it is only provisional. Making this assumption, then, let us consider the picture in question. Wittgenstein asserts that we must be able to distinguish, within a picture, a part of the picture which corresponds to each object; each such part he calls an 'element' of the picture (2.13). In our picture, the elements will be spatially separate, but this need not always be so; for example, one could picture the fact that the cat is grey by painting a picture of a grey cat. However, this is a detail; what matters is that the picture has elements, and these elements correspond to objects. Wittgenstein puts this by saying that they 'are the representatives of' (*vertreten*) objects (2.131).

We have assumed that the cat and the mat are both objects; this means that the picture has at least two elements. Let us call the element that is the representative of the cat the 'cat-representative' and the element that is the representative of the mat the 'mat-representative'. Does the picture contain any other element? It is, after all, a picture which means that a cat is *on* a mat; is there, then, an element in the picture that corresponds to the relation of being on? Wittgenstein would deny this. He would say instead: *that* the cat-representative is in a certain relation to the mat-representative 'represents that' (*stellt vor*; a form of the verb *vorstellen*) the cat and the mat are related to each other in the same way (cf. 2.15). This is why he says (2.141) that a picture is a fact; that what constitutes a picture is that its elements are related to each other in a certain way (2.14). We have just supposed that the picture which is being discussed is a true one; but it would make no difference to the picture *as a picture* if the cat were really beside the mat. By their relations, the cat-representative and mat-representative would still 'represent that' the cat is on the mat.

It is important to be quite clear about what Wittgenstein is saying here, which means being quite clear about the difference between 'to be a representative of' (*vertreten*) and 'to represent' (*vorstellen*). An element of a picture is a representative of an object; the picture itself is not. One may ask, why not? The answer is that a picture is a picture of a fact, and a fact is not an object; a fact is objects in certain relations. What the picture represents is *that* objects are related to each other in the same way that its elements are related. If the objects are so related, the picture is true; if not, it is false. It is perhaps unfortunate that the translators chose the word 'represent' to render *vorstellen*, since 'represents' in the *Tractatus* is so different from 'being a representative of'. 'Presents' (the word used to render *vorstellen* in 2.11) might have been better than 'represents'. However, there need be no confusion as long as we remember that, in the translation of the *Tractatus* that we are using, 'be a representative of' and 'represent' are not synonymous. *Elements* are representatives of objects; a *fact*—the fact that elements are related to each other in a certain way—represents *that* things are related to each other in the same way as the elements are.

Wittgenstein calls the connexion between the elements of a picture the 'structure' of the picture (2.15). (Compare 2.032, on the structure of a state of affairs.) He has already said (2.14) that what constitutes a picture is that its elements are related to each other in a certain way, so this could also be put by saying that what makes a picture a picture is that it has a structure. But there is more to a picture than this, as 2.1511–2.1515 emphasize. What constitutes a picture, it is stated in 2.1513, is also a 'pictorial relationship', by which is meant (2.1514) the correlations of the picture's elements with things. The point can be put in this way. For a picture to be a picture, it must be a picture (true or false) *of reality*, and it is by the pictorial relationship—the correlation of its elements with things—that the picture 'touches reality' (2.1515).

Although this is not made explicit, it seems to follow that it is *because* there is are certain correlations between a picture's elements and things that the elements are representatives of things. We now have to ask an important question, the answer to which is not immediately clear: namely, just what are these correlations? In the case of the picture that we are considering, the elements of the picture *resemble* the things of which they are the representatives; is the pictorial relation, then, one of resemblance? It is not hard to see that this cannot be so.

(a) We have hinted already, when quoting 2.182 earlier in this section, that not everything that the *Tractatus* would call a picture is a picture in the normal sense of the word, and this will be amply confirmed later (V.1) when we consider the proposition as a picture. In such cases the elements of the picture do *not* resemble the things of which they are the representatives.

(b) But even when there is such a resemblance, it does not follow that it is because of this resemblance that the element is a representative of an object. Consider again a picture which means that the cat is on the mat, and suppose this to be a rough drawing. The cat-representative element may resemble many cats, but in our picture it is the representative of just one; and it has to be explained how this is possible, i.e. *how* the element is correlated with just one cat. So what is required for A to be a representative of B? Number 2 of the *Tractatus* does not discuss this point, but the answer must surely be that A is *made to stand for B*, is *used as* a representative of B. If it is asked, 'Who makes it stand for B?' the answer is, 'The picture maker or makers—one or more of the "we" who (2.1) picture facts to ourselves'.

If we regard the pictorial relationship in this way, we can give an answer to some problems about picturing in the *Tractatus* which have puzzled philosophers.

1   A certain drawing, let us say, pictures a fact. In such a case, the elements of the picture will resemble the objects of which they are the representatives. But the objects will also resemble the elements of the picture; why should we not say, therefore, that the fact—the objects in their relations to each other—is a picture of the drawing?

2   Suppose that someone dribbles paint at random on to a piece of paper. Suppose also that the shapes that he produces resemble closely those produced by someone who has painted a picture which means that a certain cat is on a certain mat. Can one say that the first individual has also painted such a picture?

Before going further, consider what answers you would give to these questions.

---

1   The answer to the first question is that in so far as we say that the drawing is a picture, we are saying that parts of it, its 'elements', are made to

stand for things. One could have produced a picture which means that the parts of a drawing are in such and such relations by arranging objects in a certain order; but in this case the objects in their relations would constitute the picture, and the drawing would be the fact pictured.

2   The answer to the second question is 'No'. For something to be a picture, there must be in that something certain elements, which are representatives of things. To say this is to say that someone has *made* certain parts of a painting, say, *stand for* objects; and the paint-dribbler has not done this. We may, if we please, recognize various parts in the pattern of paint that he has produced on the paper, but none of these parts has any pictorial relationship with things, and so none of them is an element of a picture.

There is one further question to be asked about the pictorial relationship. We said that it is by correlations of the picture's elements with things that the picture 'touches reality' (2.1515). But what if the picture is a false one? Does it still 'touch reality'? Wittgenstein would reply that it does—and must, if it is to be a picture. Suppose, for example, that a certain picture which means that a cat is on a mat is not true; the cat, let us say, is sitting beside the mat. The elements of the picture are still correlated with objects; this is not affected by the falsity of the picture. What makes the picture false is that it represents that things are related in a certain way, when in fact they are not so related. We shall have more to say about false pictures later, but this must suffice for the moment.

Exercise

State in your own words the difference between 'to be representative of' (*vertreten*) and 'to represent' (*vorstellen*). (That)

III.2   PICTORIAL FORM AND THE SENSE OF A PICTURE

Reading: *T* 2.15–2.225

We spoke in the last section of the 'structure' of a picture; that is, of the connexion of its elements. Wittgenstein has a name for the possibility of such a structure. In 2.033 he says that 'form' is the possibility of structure. It is not therefore surprising that he should call the possibility of the structure of a picture the 'pictorial form' of the picture—literally, its 'form of depiction' (2.15). The introduction of the notion of pictorial form may seem puzzling. We know already that a picture must always have a structure; without it, a picture would not be a picture. Why, then, talk about the *possibility* of what a picture *must* have? The answer is that pictorial form is what a picture must have in common with what it depicts, i.e. with facts, reality (2.17). To see why Wittgenstein should say this, let us begin by considering what pictorial form is. Consider again a picture which means that a certain cat is on a mat, and suppose that the fact is that the cat is beside the mat. The elements of the picture do not have the same structure as the objects which are the constituents of the fact; nevertheless, they *could* have it. Conversely, it may be said that although the cat is not on the mat, it *could be* on the mat, and so Wittgenstein also says (2.151) that pictorial form is the possibility that things are related in the same way as the elements of the picture.

Why, then, does Wittgenstein say that it is pictorial form—the possibility of structure, rather than actual structure—that the picture has in common with what it depicts? The answer is: to allow for falsity. If the picture and what it depicts have their structure in common, that is, if the relations between the elements of the picture are the same as those between the objects of which they are the representatives, then the picture is a true one.

But Wittgenstein wishes to speak of pictures in general; of false pictures, as well as true ones. This is why he says that a picture and what it depicts have pictorial form in common; for although a false picture does not have the same structure as the fact it depicts, it *could* have the same structure.

A picture, then, has its pictorial form in common with what it depicts. But more than this: *unless* it had its pictorial form in common with reality, it would not be a picture at all. For a picture to be a picture, the objects of which the elements of the picture are representatives *must* be able to have the same relation to each other as the relation which holds between the picture's elements. Note that Wittgenstein says that this *must* be so (2.17); it is not something that depends on human decision, in the way that what a picture's elements are to stand for, i.e. be representatives of, depends on human decision.

But although a picture must have pictorial form in common with what it depicts, it cannot depict that form; rather, it displays it (2.172). What is being said here can be illustrated by means of our example. As we have just seen, a picture which means that a certain cat is on a mat must have something in common with what it depicts. The cat need not be on the mat; for example, the fact depicted (in this case, incorrectly depicted) may be that the cat is beside the mat. But if the picture is to mean what it does mean, it must be *possible* for the cat to be on the mat. In stating all this, we have stated the conditions that a picture with a certain meaning must satisfy; but the picture does not depict these conditions. What it depicts, correctly or incorrectly, is a certain fact, certain relations which actually hold between a cat and a mat.

In saying that a picture displays but does not depict its pictorial form, Wittgenstein is touching on an idea of fundamental importance to the *Tractatus*—the contrast between 'showing' and 'saying'. We shall discuss this idea in detail later (V.4); for the moment, however, let us continue to follow the course of Wittgenstein's exposition. In 2.201–2.221 he pursues further the topic of the way in which a picture depicts reality, and in the course of these propositions he defines the important word 'sense' (*Sinn*)—a term used, though not explained, earlier in the *Tractatus* (2.0211). He says first (2.201) that a picture depicts reality in that it 'represents' (*darstellt*) a possibility of existence and non-existence of states of affairs. This proposition needs close examination. We need to find answers to two questions;

(a) What is meant by 'representing'?

(b) What is represented?

Let us take the second question first. Wittgenstein says in *T* 2.06 that to speak of a fact is to speak of the existence and non-existence of states of affairs; it looks, then, as though we may say that a picture represents a possibility of a fact. Nor, indeed, does there seem to be anything seriously wrong in saying this, and in what follows we shall continue to do so. It must be noted, however, that the *Tractatus* does not use terms whose literal translation would be 'the possibility of a fact' or 'a possible fact'. Instead, it speaks of a possible *situation* (*Sachlage*), or the possibility of a situation (2.202, 2.203). However, in 2.11 a situation is said to be the existence and non-existence of states of affairs, so it appears that 'fact' and 'situation' are equivalent terms.

There is one qualification to be added to this. We have argued that a picture represents the possibility of a fact; but it should not be assumed that *every* picture does this. Wittgenstein would say that in some cases a picture represents the possibility, not of a fact, but of a state of affairs. (See the discussion of elementary propositions in VI.1 below.) For the moment, however, Wittgenstein is concerned with pictures of facts (cf. 2.1), and our discussion will concentrate on such pictures for the present. We shall therefore continue to speak of a picture as representing the possibility of a fact.

Let us now consider what is meant when it is said that a picture 'represents' what we have called the possibility of a fact. 'Represent' here does not mean 'be a representative of', in the way in which an element of a picture is a representative of an object; a fact, whether actual or possible, is not an object. *Darstellt* (like *vorstellt* in 2.15; cf. III.1) would be better translated as 'presents', and indeed the two words seem here to have the same meaning. It is true that in 2.15 what is said to be represented is that things are related in a certain way (namely, as the elements of the picture are related to each other), whereas in 2.201 what is said to be represented is the possibility of a fact. But there is no essential difference between what is said in these two propositions. In representing, e.g., that the cat is on the mat, the picture represents only a possible fact; really, the cat may not be on the mat. It is true that the picture depicts a fact, which makes it true or false, but it does not represent that fact. What it represents, or presents, is a possible fact.

Wittgenstein says (2.221) that what a picture represents is its sense (*Sinn*). We have argued that a picture represents a possible fact, and so we may infer that the sense of a picture is a possible fact. The word *Sinn* can also be translated as 'meaning', so it could be said that the meaning of a picture is a possible fact. We shall need to consider the term *Sinn*, and the nature of meaning in general, in greater detail when we discuss Wittgenstein's views about the proposition in the next section; for the moment, we need only note that the meaning or sense of a picture is very different from the meaning of the elements of a picture. A picture represents, or presents, possible facts; its elements are the representatives of, are correlated with, objects.

The introduction of the term 'sense' enables Wittgenstein to be more precise about the nature of truth. He has said (2.21) that a picture either agrees with reality or does not; it is correct or incorrect, true or false. From this one might infer that a picture is true or false in so far as it agrees or disagrees with reality. In 2.222, however, it is said that what constitutes a picture's truth or falsity is the agreement or disagreement of its *sense* with reality. There is no inconsistency here. Wittgenstein is simply saying that what agrees or disagrees with reality is not the picture considered, say, as a drawing; what agrees or disagrees with reality is what it *means*. We shall have more to say later about Wittgenstein's views on falsity (VI.5).

Exercises

1   What is the difference between a 'pictorial relationship' and 'pictorial form'? (If you are unclear about the answer, refer back to the discussion of 2.1514 in III.1 above.)
2   State in your own words the difference between 'depicting' and 'representing'.

III.3   LOGICAL FORM

Reading: *T* 2.18–2.2

So far we have restricted our attention to spatial pictures, and it is now time to widen our account. Wittgenstein says (2.182) that not every picture is a spatial picture, and in 2.171 he says that there are spatial pictures, coloured pictures, 'etc.'. What is covered by the 'etc.'? A way of answering this is to ask what something *must* be if it is to be a picture in Wittgenstein's sense; then we shall know, of possible candidates for the title of 'picture', which are satisfactory and which are not. What we have seen already is sufficient for us to construct an answer to this question.

Exercise

Before going further, attempt your own answer to the question, 'What must something be if it is to be a picture in Wittgenstein's sense?'

We know from 2.171 and 2.182 that a picture may (but need not) be spatial, that it may (but need not) be coloured. We know, however, that a picture *must have* certain elements, each of which is the representative of (i.e. is correlated with) an object. We know also that the elements *must have* certain relations to each other, such that they 'represent' that objects are related in the same way. This implies that a picture must have its 'pictorial form' in common with reality; that is, that it must be such that things *can* be related to each other as the elements of the picture are. These conditions appear to be sufficient.

It is not, for example, necessary that the elements of a picture shall resemble the objects of which they are the representatives. Since we have hitherto been taking as examples what would be counted as pictures in the normal sense of the word, we have assumed that there is such a resemblance. Certainly, there *can* be such a resemblance, as is implied by the reference to hieroglyphics in 4.016, but Wittgenstein nowhere says that such resemblance is necessary. As this is so, it is possible to picture the fact that the cat is on the mat by (e.g.) correlating with the cat the word 'cat', and with the mat the word 'mat'. A true picture of the fact would be:

cat

mat

and false pictures would be:

mat
              cat   mat   mat   cat
cat

Nor is it necessary for the picture to be in any way spatial; we can, for example, produce a musical picture, and this would serve as an example of what is covered by the 'etc.' in 2.171. Asked, say, 'What are the relations of pitch between the first two notes of Beethoven's fifth symphony?', I can answer by *singing* the notes. The first note sung is the representative of the first note of the symphony, and the second note sung is the representative of the second note of the symphony. The fact that the notes sung have certain pitch relations represents that the notes of the symphony are related in the same way. If the first note sung has the same pitch as the second, the picture is a true one; if not, it is false.

But although there are pictures of various kinds—spatial, coloured, musical—Wittgenstein says that every picture, of whatever kind, must have something in common with reality, and this something is 'logical form' (2.18). He adds (2.181) that a picture whose pictorial form is logical form may be called a 'logical picture'. This may appear puzzling. We make spatial pictures by (e.g.) drawing things, or constructing three-dimensional models; we make musical pictures by producing sounds. But how do we make a logical picture? This question rests on a misunderstanding. There is no especial way of making a logical picture; every picture is *at the same time* a logical picture (2.182), so in producing, say, a spatial picture we are also producing a logical picture.

But what is a logical picture? What is that 'logical form' which is closely related to such a picture? The notion of logical form is not easily explained, and indeed Wittgenstein says (4.121) that logical form cannot be expressed by means of language. However, one can at any rate try to give some indication of what is meant. First, in saying that every picture must have logical form in common with reality, Wittgenstein does not mean that it must have a common structure. If he did, then (as was seen in III.1) he must mean that every picture, in so far as it is a logical picture, is true; but the assertion (2.19) that logical pictures can depict the world implies that they can be true or false. (As will be seen later (VI.3), a picture which is always true would be a tautology, and would not depict the world.) Logical form, then, like pictorial form, must be a *possibility* of structure, and that this is so is implied by the fact that Wittgenstein runs the notions together in using the term 'logico-pictorial form' (2.2).

Logical form, then, is a possibility of structure; but of what structure? Perhaps the simplest answer is: structure of the most abstract, the most general kind, such that it can be shared by spatial pictures, musical pictures, and in short by pictures of all kinds. To make what is meant clearer, suppose that someone pictures a certain fact by producing a drawing which has two elements, one of which is higher than the other. Suppose also that someone else pictures a certain fact by producing a musical picture, whose two elements are two notes played in succession, the one being louder than the other. These are pictures of different kinds, yet each may be said to have the same structure, in that the relations of higher than and louder than are both what are called transitive asymmetrical relations. (Transitive, in that if e.g. *A* is higher than *B* and *B* is higher than *C*, then *A* is higher than *C*; asymmetrical, in that if *A* is higher than *B*, *B* is not higher than *A*.) To speak of logical form, then, is to speak of the possibility of structure of this kind, and a logical picture is one that represents a possible fact the relations between whose objects are of this very general kind.

The notion of a logical picture is used both in Wittgenstein's account of the nature of a thought (*T* 3) and of a proposition (4.03), topics that will concern us in later sections. But before we go on to discuss these topics, it may be useful to give a short summary of Wittgenstein's views about pictures, as these have been presented so far. We can do this in the form of a comment on the assertion with which our inquiry began, 'We picture facts to ourselves'. Let us ask what it is to picture a fact, e.g. by making a drawing.

Exercise

In about 300 words, attempt your own answer to this question.

To picture a fact by making a drawing, it is not enough just to make a drawing. The drawing must mean something, in the sense that we can say of it, 'That's a true picture' or 'That's false'. A drawing of this kind must contain 'elements', parts of the drawing each of which is the representative of an object. That is, we must correlate the elements of the picture with objects; we must lay it down that such and such a part of the drawing is to be a representative of such and such an object. Further, not only must there be correlations between the elements of the picture and objects, but the picture

Present

must also mean that these objects stand in certain relations. It does this by virtue of the fact that its elements are in certain relations to each other, and what it means is that the objects are related as its elements are related. The picture's meaning or 'sense' is a *possible* fact; what makes a picture true or false is a *fact*, and it is this fact which the picture depicts. A picture must therefore be such that its elements can have the same relations to each other as the objects of which they are the representatives. But the picture does not depict this, it displays it.

You may ask, 'Where is all this leading us?' The answer is that it is leading us to Wittgenstein's account of the proposition, which is at the heart of the *Tractatus*, and to which we now turn.

IV   PROPOSITIONS

IV.1   THE THOUGHT AS A PICTURE

Reading: *T* 3–3.02, 3.1

Number 3 of the *Tractatus* begins with the assertion that 'A logical picture of facts is a thought' (*Gedanke*). This at once faces us with a problem: namely, how the word 'thought' is to be understood here. Since a thought is said to be a picture of facts, we may expect it to have some connexion with truth and falsity, and indeed 3.01 speaks of 'true thoughts'. The problem is, what is the precise connexion between a thought and truth or falsity? Is a thought the *thinking that* something is true or false, or is it *that which* is regarded as true or false, i.e. that of which truth or falsity is predicated? Frege took the word 'thought' in the latter sense. A thought, he said, is 'something for which the question of truth arises', and so we can ascribe both truth and falsity to thoughts. (See 'The Thought: A logical Inquiry', trans. by A. M. and Marcelle Quinton, in P. F. Strawson (ed.), *Philosophical Logic*, Oxford, 1967, p. 20.) A thought, then, is not for him 'the subjective performance of thinking' ('On Sense and Reference', in *Philosophical Writings of Gottlob Frege*, trans. by P. Geach and M. Black, Oxford, 1952, p. 62n.). Rather, it is in effect what is commonly called a proposition or judgement—that is, that which is true or false ('The Thought', p. 20, n.1). Is this how Wittgenstein took the word? 3.01, with its reference to 'true thoughts', may suggest that it is; however, the German text of 3.02 tells against this interpretation. Pears and McGuinness render the first sentence of 3.02 as 'A thought contains the possibility of the situation of which it is the thought'. The German for the last phrase, however, is 'die es denkt'—literally, 'which it thinks'. This hardly suggests that a thought is a proposition; propositions do not think. It is true that, strictly speaking, thoughts do not think either; however, we may take the phrase to imply that a thought is an act of thinking, i.e. what Frege called a 'subjective performance'.

This is confirmed by a letter to Russell (19.8.19), in which Wittgenstein answered questions on the *Tractatus*. Russell had asked whether a thought consists of words. Wittgenstein's reply was, 'No! But of psychical constituents that have the same sort of relation to reality as words' (*N*, p. 130). Now a proposition, in the sense of that which is true or false, is not made up of psychical constituents. For example, two people can agree that a proposition is true; but if a proposition consisted of psychical constituents, this could not be so—two separate propositions would be involved, each consisting of constituents of one individual's psyche or mind. Frege in effect made this point when he said ('Sense and Reference', loc. cit.) that a thought, in his sense of the term, is the objective content of thinking, 'which is capable of being the common property of several thinkers'. But a psychical constituent is not common property.

This interpretation of the meaning of 'thought' in the *Tractatus* is not without its difficulties. It was pointed out earlier that in 3.01 Wittgenstein speaks of 'true thoughts'. But how can an act of thinking be called 'true'? An act can in principle be dated; what is true or false, cannot. It makes no sense to say, 'Pythagoras' theorem has been true for over 2,000 years', though it does make sense to say that Pythagoras first proved his theorem over 2,000 years ago. Perhaps the answer to this is that to think (as this term is used in the *Tractatus*) is to think *that* something is the case. One can say, 'In thinking that so and so, he was right', from which it is a short step to saying, 'His thought was a true one'.

So far, we have seen that when the *Tractatus* speaks of a 'thought', what is meant is a *thinking that* so and so is the case. But this raises a further problem: the problem of what it is to think that so and so is the case. The letter to Russell quoted earlier, with its reference to a thought as consisting of psychical constituents, may be taken as implying that, in the *Tractatus*, thinking is regarded as something non-physical, immaterial. To see whether or not this is so, we need to give some consideration to what the *Tractatus* says about the proposition (*Satz*). In 3.1 it is said that in a proposition, a thought finds an expression which is perceptible by the senses. We shall explore the nature of the proposition more thoroughly in later sections; for the moment, let us simply ask whether 3.1 means that words, say, are the physical expression of something that is essentially non-physical. A view of this kind is not without precedent. From Descartes onwards, many philosophers have supposed that thinking is the activity of an immaterial substance, and that the activities of such a substance can be perceived directly only by the agent; all that others can perceive is the physical behaviour which is the expression of (which often means, is caused by) the activities of this substance. In his later philosophy, Wittgenstein attacked this view severely. In this case, however, he does not seem to be attacking his earlier views. Looking again at the letter to Russell, we note that the psychical constituents are said to have the same kind of relation to reality as words. There is no suggestion that the former are more fundamental; no suggestion that the real business of thinking is carried on by means of psychical constituents, and that words are merely their external form.

In sum, there is no good reason to believe that, in the *Tractatus*, a thought is something that is essentially immaterial. But we are still left with the question, what *is* it? The answer is not given until *T* 4, where it is said that 'a thought is a proposition with a sense'. To understand this, we need to learn more about Wittgenstein's concept of a proposition, and this will be our concern in the next two sub-sections.

IV.2 PROPOSITIONS, PROPOSITIONAL SIGNS AND PROJECTION

Reading: *T* 3.1–3.14

Before we discuss the meaning that Wittgenstein gives to the term 'proposition' in the *Tractatus*, it will be useful to prepare the ground by saying something about the sense which philosophers commonly give to the word. We have already said in IV.1 that a proposition is commonly regarded as that which is true or false. It is also usual to ascribe to propositions the following features:

(i) A proposition is distinguishable from the sentence that expresses it. This is because the same truth or falsehood, i.e. the same proposition, can be expressed by different sentences, whether in different languages or in one and the same language.

(ii) A proposition is what is believed, conjectured, presupposed, entertained, etc. The connexion between this feature and the concept of a proposition as that which is true or false is that when, for example, one believes that so and so is the case, what one believes is true or false.

(iii) Although a proposition can always in principle be expressed publicly— e.g. in words—it need not be so expressed. One can, for example, entertain the proposition that so and so is the case, and not actually say or write anything.

We can now ask how a proposition, as this term is understood in the *Tractatus*, is related to a proposition of the kind that has just been described. There are two clear differences.

Exercise

Before going further, consider what these differences are.

(a) We have already seen that a proposition, as the *Tractatus* understands it, is the perceptible expression of a thought (3.1). These perceptible expressions are usually spoken or written signs (3.11), of which words are the most common. For the sake of brevity, we will assume henceforth that a thought is expressed only in verbal form. Now we know that, in the *Tractatus*, a 'thought' is not what is commonly called a proposition. It is not that which is true or false, but is an act of thinking, though it is an act which (to borrow a term from J. L. Austin) has a truth-falsity dimension. It is clear from this that what the *Tractatus* calls a proposition, namely the perceptible expression of a thought, will also differ from a proposition in the usual sense of the term.

(b) A proposition, in the usual sense of the term, need not be expressed in words, or indeed in any perceptible form. But a proposition, as the *Tractatus* understands it, *is* an expression of a thought (an act of thinking) in sensibly perceptible form.

We have already said that this does not mean that (e.g.) an inaudible act of thought is expressed by audible words, and it is now time to ask what it does mean. As Wittgenstein has said (*T* 3) that a thought is a picture, it might be supposed that he would give an account of the proposition—the perceptible expression of a thought—in terms of pictures and picturing. He does indeed say that a proposition is a picture of reality, but not until 4.01—that is, several pages after his account of the proposition has begun in 3.1. Between 3.1 and 4.01 there is hardly any express reference to pictures; instead, Wittgenstein first explains the nature of the proposition in terms of what he calls the 'propositional sign' and 'projection'. The course of his exposition is as follows: he first explains the nature of the proposition, in terms of the propositional sign and of projection (3.11–3.14), and he next refines upon the notion of a propositional sign (3.14–3.144). We will follow this order, reserving the argument of 3.14–3.144 for the next sub-section. To anticipate: we shall find that although the terms 'propositional sign' and 'projection' are new to the *Tractatus*, the ideas that lie behind them are not. There may be almost no explicit reference to picturing between 3.1 and 4.01, but the idea of picturing is certainly present.

A proposition, Wittgenstein says, is a propositional sign in its projective relation to the world (3.12). He explains what he means by a propositional sign by saying (3.12) that it is the sign with which we express a thought; it is (3.11) the 'perceptible sign'—spoken or written, etc.—of a proposition, and its elements are words (3.14). This may seem puzzling. Wittgenstein has said in 3.1 that a thought finds expression in a proposition; he now says that a propositional sign is the sign with which we express a thought. A proposition and a propositional sign, then, seem to be described in much the same way; yet the definition of a proposition in 3.12, quoted at the beginning of this paragraph, implies that they are different. The solution to the problem is that proposition and propositional sign are not different kinds of thing. Like the proposition, the propositional sign consists of words (3.14); the difference between them is that a proposition is a propositional sign used in a certain way (3.11).

So far, then, we have seen that for Wittgenstein a proposition is an act, the uttering something true or false, and to say something true or false is to use a propositional sign—words—in a certain way. But in what way, precisely? On the basis of what he has said about pictures, we might expect Wittgenstein to say that a proposition is a propositional sign used to represent a possible situation and to depict the world. This, indeed, seems broadly to be what he means, but the term that he uses to express his meaning here is the term 'projection'. The propositional sign, he says, is used as a projection of a possible situation (3.11). Outside 3.11–3.13, the term 'projection' makes only one further appearance in the *Tractatus* (4.0141), but it will be worth while to examine carefully what Wittgenstein means by it—both for the light that it throws on his views about the proposition, and because the concept of projection recurs in some of his later works (cf. Lectures, 1930–33, in G. E. Moore, *Philosophical Papers*, London, 1959, pp. 263–4, 270; *The Blue Book*, p. 53; *Zettel*, paras. 137, 290–91).

We saw just now that 3.11 states that the propositional sign is used as a projection of a possible situation. As in 2.202–3 (cf. III.2 above), 'possible situation' seems to mean 'possible fact'. Wittgenstein does not explain what he means by 'projection' here, but there is every likelihood that his meaning of the term is related to its meaning in projective geometry. A detailed account of projection of this kind is unnecessary here, but we must mention some of the basic ideas involved. In 3.11 Wittgenstein says that something is a 'projection of' something, and in 4.0141 he says that something is 'projected into' something. What do these terms mean, in their normal use in projective geometry? To answer these questions, let us suppose that there is a plane surface, P, and that on P there are three dots. Choose any point outside the plane of P; call this point the 'origin', O. Construct lines that pass through the points on P and touch a second plane, Q. These lines may be called 'lines of projection'. The three points that are formed on Q are called a 'projection of' the original figure; the original figure is also said to be 'projected into' the new one.

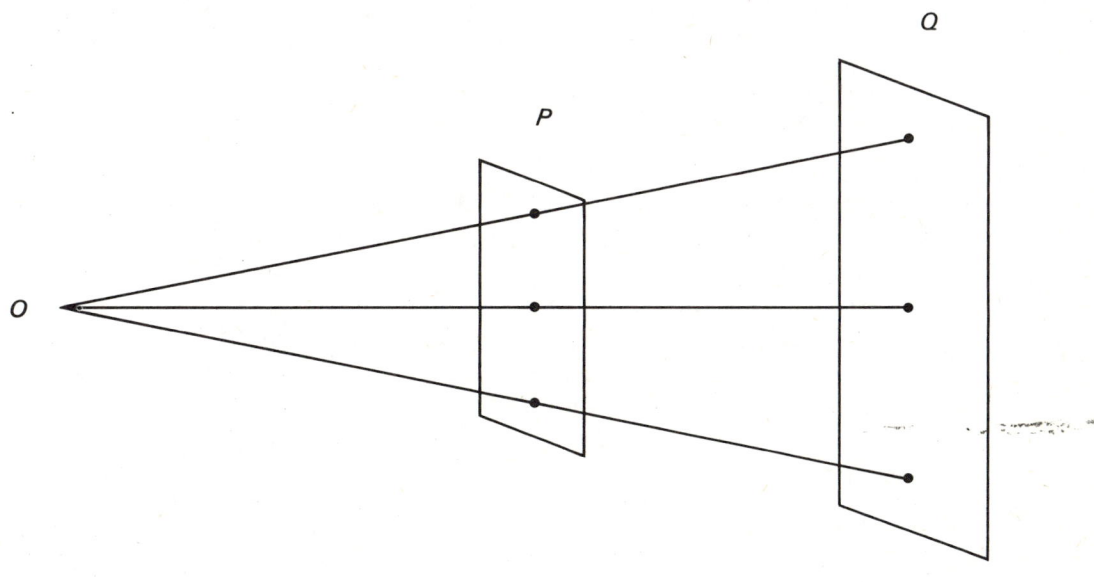

This is the principle of the slide projector or cinema projector; *P* would here be a slide or a frame of a film, *O* a source of light, and *Q* a screen. We could also (and here we make contact with what was said earlier about pictures) regard the figure on *P* as a projection of the figure on *Q*. In this case, the points on *Q* in our diagram would represent something to be copied, and *P* might be a piece of transparent material on which an artist, whose eye is at *O*, draws. It should perhaps be added that our diagram is an example of just one sort of plane projection, known as 'central projection'; if the lines going through the points on *P* had been parallel, we should have had an example of another type of projection, known as 'parallel projection'. It may be said, however, that in all cases of projection lines are drawn through all the points on one figure, *F*, and produce on another surface a new figure, *G*, each point of which corresponds to a point of *F*.

Now, how is all this to be applied to propositions and propositional signs? Clearly, in explaining the nature of a proposition in terms of what he calls 'projection', Wittgenstein is using a metaphor. He has said that a proposition is a propositional sign used as a projection of a possible situation. Now, as we know, a propositional sign consists of words; let us suppose, for the sake of argument, that such a sign consists of spoken words. We cannot say, in this case, that lines of projection literally lead to or pass through these words. How, then, are we to take what Wittgenstein says about projection here? What analogies are there between geometrical projection and the nature of the proposition? Let us consider this question in the light of our diagram.

If we suppose Wittgenstein to have pictures in mind, then (since *P* in our diagram is a picture of *Q*) the figure on *P* will correspond to the propositional sign, and the figure on *Q* will correspond to the possible fact of which the propositional sign is the projection. It will be worth while to reflect for a moment on why the figure on *Q* has to stand for a *possible* fact.

Why do you think that this is so?

As we have seen, Wittgenstein is trying to explain the nature of the proposition, i.e. the saying something true or false. Now, if the points on *Q* were regarded as standing for an actual fact, then the diagram would represent a *true* proposition. (E.g. the points on *Q* might be a pattern on a wall, and the points on *P* an artist's representation—a faithful representation—of that pattern.) But it is necessary to allow for the possibility of a proposition's being false, and for this reason the figure on *Q* must be regarded as standing, not for an actual, but for a possible fact.

What, then, are we to make of what we have called 'lines of projection'? To find an analogy to these in the philosophy of the *Tractatus*, let us refer back to our discussion of the elements of a picture in III.1. We noted there that Wittgenstein says that a picture 'touches reality' (2.1515); it does so, we may now add, by means of what he calls certain 'feelers' of its elements. This is clearly metaphorical, and in fact Wittgenstein indicates that by 'feelers' he means the correlations of the picture's elements with things, i.e. the 'pictorial relationship' (2.1514). Applying this to projection and the proposition, it seems that the analogue to 'lines of projection' in the case of the proposition will be the correlations of the elements of the propositional sign with things.

But this is not all; the use of a propositional sign of which Wittgenstein is speaking is analogous to geometrical projection in another respect. Consider our diagram again, and suppose that the figure on *P* is a projection of the figure on *Q*. It will be noticed that points *in a certain order* on *P* are a projection of points *in a certain order* on *Q*. We might put this in the language of the *Tractatus* by saying that one fact is a projection of another. To get closer to the idea of the propositional sign as a projection, suppose now that the figure on *P* is a picture in Wittgenstein's sense, i.e. it is used to say something true or false. In this case, the dots on *P* may be regarded as a projection of the possible fact represented by the three dots on *Q*. (This possible fact would be the possibility that three dots, of which the dots on *P* are representatives, are related as the dots on *P* are related.) This could be expressed by saying that one fact—i.e. the dots on the picture *P*—is a projection of another (possible) fact. Wittgenstein argues that something similar holds in the case of the propositional sign, as we shall now see.

IV.3 THE PROPOSITIONAL SIGN AS A FACT

Reading: *T* 3.14–3.144

Just as Wittgenstein says (2.141; cf. III.1) that a picture is a fact, so too he says that a propositional sign is a fact (3.14). What constitutes a propositional sign, he says, is that in it its elements, words, stand in a determinate relation to each other. He is emphatic about this, on the grounds that it is obscured by the usual forms of expression in writing or print (3.143). When we print a proposition, for example, it is not evident from the printed page that there is an essential difference between a propositional sign and a word. Yet there is such a difference. A propositional sign is not just a string of words, it is a fact—the fact that the words which make it up stand in certain relations to each other. This can be seen quite clearly, Wittgenstein says (3.1431), if we imagine a propositional sign consisting of spatial objects instead of written signs; the spatial arrangement of these objects will express the sense of the proposition. We may illustrate this by an example from Wittgenstein's *Notes on Logic* of 1913 (*N*, p. 98), where he says that the fact that this inkpot is on this table may express the fact that I am sitting in this chair.

In the *Tractatus*, Wittgenstein makes his point by an example from logical symbolism. Consider the propositional sign '*aRb*' which, in logic, is a standard way of saying that *a* stands in the relation R to *b*. It would, according to Wittgenstein, be wrong to say that the sign '*aRb*' says that *a* stands in the relation R to *b*. His point here is that this would imply that the propositional sign is just a string of signs. What we ought to say is: *that* '*a*' stands in a certain relation to '*b*' says *that a* stands in the relation R to *b* (3.1432).

This proposition has given rise to a great deal of discussion. It has been thought that Wittgenstein means that the use of the sign for a relation, 'R', is logically wrong. That is, in order to say that *a* stands in the relation R to *b* we ought not to write '*aRb*', but should simply write the signs *a* and *b* in a certain relation to each other; that these signs are in this relation will say that *a* stands in the relation R to *b*. Similarly, Wittgenstein has been thought to mean that if we wish to say that the cat is on the mat we ought not to write 'The cat is on the mat', but should write simply (e.g.)

cat

mat.

In III.3 it was suggested that the possible fact that the cat is on the mat *could* be represented in this way; the present suggestion is that it is Wittgenstein's view that it *ought* to be represented in this way—or, if not in precisely this way, in some other way which does not involve a word or phrase for the relation of being on.

Such a view would involve a serious difficulty: namely, the difficulty of expressing, without special words for relations, the enormous variety of relations that there are. Perhaps a way could be found out of the difficulty; however, there is in any case good reason to believe that the view just described was not Wittgenstein's. In 4.012 he refers again to a proposition of the form '*a*R*b*', and says that in this case the sign is 'obviously a likeness of what is signified'. But if the interpretation of 3.1432 suggested above is correct, '*a*R*b*' would decidedly not be a likeness of what is signified. What, then, does 3.1432 mean? Consider again the propositional sign '*a*R*b*', which says that *a* stands in the relation R to *b*. Note what it says: that *a* stands in the relation R to *b*, and not that *b* stands in the relation R to *a*. Now, by virtue of what does it say what it says? Wittgenstein can be taken as saying that it does so by virtue of *the fact that a* stands to the left of *b* in the printed or written complex sign. There is no suggestion, however, that the use of the sign 'R' is wrong.

Let us now return to the topics of the propositional sign and of projection. Summing up the results of our inquiries in this and the last sub-section, we can say that when Wittgenstein argues that a proposition is a propositional sign used as a projection of a possible situation, he means:

(i) The elements of a propositional sign—words—are correlated with things, i.e. they are the representatives of things.

(ii) The fact that these elements are in certain relations represents a possible fact; namely, that the things of which they are the representatives are related to each other in the same way.

We saw early in this sub-section that Wittgenstein speaks of the 'sense' of a proposition (3.1431), and indeed there are several occasions on which he does this (3.144, 3.3, 4.021, 4.022, 4.031). We have already spoken in III.2 of the sense of a picture, and we argued there that when Wittgenstein speaks of a picture as representing, or presenting, its sense (2.221) he means that it presents a possible fact. It may be inferred that the sense of a proposition is a possible fact, namely the possible fact of which a certain propositional sign is a projection.

It will now be clear that, as was suggested earlier, what is said about the proposition in terms of projection is an application to propositions of what was previously said about pictures. But this does not mean that our investigations of the last two sub-sections have been superfluous. First, they have been an inquiry into the nature *of the proposition*, and that is a topic on which the parts of the *Tractatus* discussed in Section III did not touch. Second, they have given us a more complete understanding of what is meant by picturing in the *Tractatus*—not that our inquiry into that topic is yet at an end (cf. V.1 below).

We shall have more to say in later sections about the account of the proposition given in the *Tractatus*, but we have already seen enough to be able to answer a question asked at the end of IV.1. This is the question that preceded our inquiry into the nature of the proposition: namely, what is a thought? That is, what *is* it of which a proposition is the perceptible expression? We said in IV.1 that a thought need not be immaterial, but we did not try to say what it is; we were not at that stage able to say even roughly what is meant by the statement (*T* 4) that a thought is 'a proposition with a sense' (literally, 'the significant proposition'—*der sinnvolle*

Satz). What we have seen in IV.2 and IV.3 seems to indicate that this means, not that a thought is something behind the spoken or written words which are its public manifestation, but rather that the thought *is* the speaking or writing etc. of what is true or false. Perhaps, however, this is over-simplified; we must remember Wittgenstein's letter to Russell quoted in IV.1, in which it is said that the constituents of a thought are not words, but are psychical. Can we reconcile these apparently contradictory points of view? Perhaps we can. Wittgenstein's view may be that although a thought need not be made public, such a thought must (like every thought) be a 'logical picture' (*T* 3). In such a case, he may say, we are making use of something that corresponds to a propositional sign—for example, silent speech. We may call such a sign a 'quasi-propositional sign'; unlike a propositional sign properly so-called, it is not perceptible by others, but, like a propositional sign, it has elements and a structure. The important point is that to think is to use signs—either propositional signs, of the kind discussed in the *Tractatus*, or quasi-propositional signs of the kind that we have just suggested—and to use them in the way that has been described in the last two sub-sections, i.e. as a projection of a possible situation. In thinking, then, the signs used may be *either* public *or* private. The view of the *Tractatus* seems to be that it would be wrong to say that thinking is really a private process, of which words are merely the external manifestation, but that it would be equally wrong to say that all thinking is public, in that it is the use of perceptible signs.

We have been discussing Wittgenstein's concept of a proposition; to conclude this sub-section, let us think again about the way in which the term 'proposition' is commonly used by philosophers. It was pointed out early in IV.2 that in this usage, 'proposition' means that which is true or false, and that a proposition is also distinguished from the sentence that expresses it. Wittgenstein has been able to preserve this distinction. For him, what is true or false is not the sentence—the propositional sign—as such; it is the propositional sign in its projective relation to the world, the propositional sign *as used* to represent a possible fact. He is also able to put in his own terms what the traditional theory of the proposition states by saying that various sentences can express the same proposition; he can say that different propositional signs can be projections of the same possible fact.

IV.4 NAME, OBJECT AND MEANING

Reading: *T* 3.2–3.22

You may have gained the impression that the *Tractatus*, as so far examined, falls into two parts which are only loosely related to each other. The first part was concerned with facts, states of affairs and objects, and the second has been concerned with pictures and propositions. The only connexion between the two that has been seen so far has been that pictures (and these include logical pictures, thoughts) are pictures of facts. However, the opportunity to bring together more closely the two parts is provided by the next topic to be discussed in the *Tractatus*—names and their meaning.

In 2.13 it was stated that the elements of a picture correspond to objects, of which they are the representatives. We have seen recently that a propositional sign, too, has elements, and that these are words (3.14). Wittgenstein now says (3.2) that it is possible for a thought to be expressed in a proposition in such a way that the elements of the propositional sign correspond to the objects of the thought. He calls such elements of a propositional sign 'simple signs' (3.201); when they are employed in propositions he calls them 'names' (3.202). The distinction may seem a fine one, but it is clear enough. Certain elements of a propositional sign, considered simply as words with certain relations to each other, are called 'simple signs'. When a propositional sign is *used*, used as a projection of a possible situation, these same elements are called 'names'.

The way in which 3.2 is expressed—'In a proposition a thought can be expressed in such a way that elements of the propositional sign correspond to the objects of the thought'—suggests that it is possible for an element of a propositional sign *not* to correspond to an object. This is confirmed by the third paragraph of 3.24, where it is said that a 'propositional element' may signify a complex. A proposition which contains such a propositional element can be analysed; that is, it is possible to replace the propositional element in question by elements each of which corresponds to an object of the thought which the proposition expresses. This is why it is said (3.201) that a proposition the elements of whose propositional sign correspond to the objects of the thought—that is, a proposition whose elements are 'simple signs'—is to be called 'completely analysed'.

It has already been seen (II.5.1) that Wittgenstein thinks that it is not the philosopher's business to give examples of objects, so it is clear that we are not to expect from him any examples of completely analysed propositions. Nevertheless, what he says about simple signs and names is valuable for the light that it throws both on his views about meaning, and on his reasons for believing that there *are* objects. The first of these topics will be discussed in this sub-section, the second in IV.5.

The elements of a picture, we saw, are the representatives of objects (III.1; *T* 2.131); correspondingly, in a proposition a name is the representative of an object (3.22). Wittgenstein adds (3.203) that a name *means* an object; the object is its meaning (*Bedeutung*). This is an important statement, which needs careful consideration. Frege had distinguished between what it is common to translate as the 'reference' (*Bedeutung*) and the 'sense' (*Sinn*) of names. The 'reference' of a name is the object for which the name stands; the 'sense' is what it means. It is necessary to distinguish between the two, Frege argued, because names can have the same reference but different senses ('On Sense and Reference', *Philosophical Writings*, trans. Geach and Black, p. 57). For example, 'Morning Star' and 'Evening Star' have different senses, but the same reference, the planet Venus; so to identify sense and reference, to say that the meaning of a name is the object for which it stands, must be wrong.

It might not be thought obvious that 'Morning Star' and 'Evening Star' do have different senses. Why should we not say that the two names not only stand for the same planet, but also mean the same? Frege would reply as follows ('Function and Concept'), op. cit., p. 29). Take the two propositions: 'The Morning Star is a planet with a shorter period of revolution than the earth' and 'The Evening Star is a planet with a shorter period of revolution than the earth'. Now, someone might think that one of these propositions is true and the other false, which shows that the two propositions do not have the same sense. Sense and reference, then, are not the same; a proper name expresses its sense, but stands for or designates its reference ('On Sense and Reference', op. cit., p. 61).

Now let us compare Wittgenstein's views in the *Tractatus*. A name, we have seen, is the representative of an object, which makes it clear that the object is the reference of the name. But it is also Wittgenstein's view—contrary to that of Frege—that the object is the *meaning* of the name. There is, incidentally, no doubt about the accuracy of this translation of *Bedeutung*, which is also that given in the Ogden and Ramsey version, and which was not queried by Wittgenstein. As supporting evidence, we may cite *T* 3.263. This proposition (which will be discussed in detail below, IV.6) says that certain *Bedeutungen* can be explained by 'elucidations'. Now, it is clear that what one elucidates or clarifies are meanings.

In the *Tractatus*, then, the meaning of a name is what Frege would call the 'reference' of the name; it is what the name stands for. The *Tractatus* gives no indication of Wittgenstein's reasons for thinking that this is so. However, there are some obvious points in favour of the view. It explains how different people can mean the same thing; it also avoids the idea that the meaning of a name is some kind of shadowy entity. If one takes the view that the meaning of a name is the reference of that name, then the meaning of the name 'Venus' (assuming that it is a name) is no more shadowy than the planet itself.

We have spoken so far about meaning, *Bedeutung*, in the *Tractatus*; but it should also be noted that what Wittgenstein means by *Sinn*, sense, is different from what Frege means. For Frege, a name has both reference and sense; for Wittgenstein, on the other hand, sense belongs to propositions only and not to names (3.3; cf. 3.144). This is because, in the *Tractatus*, to speak of 'sense' is to speak of a possible fact (III.2, IV.3), and a name cannot present a possible fact; a name says nothing, it makes no assertion, true or false.

It was suggested earlier that a picture (which, it will be recalled, presents a possible fact) can be said to have a meaning (III.2, III.3), and indeed it is usual to speak of Wittgenstein's 'picture theory of meaning'. This can be justified, but it can also mislead. As far as names are concerned, the *Tractatus* offers what is commonly called a 'denotation theory' of meaning; the meaning of a name is its reference, is what it 'denotes'. Wittgenstein's 'picture theory of meaning' is not a theory about the meaning of names, it is a theory about the meaning of propositions.

There is one further point to be made here about names in the *Tractatus*. After saying in 3.203 that a name means an object, and that the object is its meaning, Wittgenstein adds in parenthesis that '*A*' is the same sign as '*A*'. The point of this remark is not immediately clear from the context; however, in his 1913 *Notes on Logic* (*N*, p. 104) Wittgenstein explained a similar remark by saying that names are not things but classes. What he is saying can be expressed by means of a distinction that is commonly drawn between words as types and words as tokens. This distinction may be explained as follows. In the second sentence of this paragraph, the word 'Wittgenstein' occurs; the word 'Wittgenstein' also occurs in the third sentence. Now, do we have here two separate words, or two instances of one and the same word? The answer to this question depends on how the term 'word' is understood. If we say that there are two separate words here, we are regarding the word 'Wittgenstein' as a *token*. If, on the other hand, we say that there are in these sentences two instances of the word 'Wittgenstein', we are regarding the word as a *type*. Similarly, if we say of a book that it consists of 90,000 words we are regarding words as tokens, but if we make an index of all the words used in a book we shall regard words as types. Now it is clear, from *T* 3.203 and the corresponding passage in the *Notes on Logic*, that in the *Tractatus* a name is regarded as a type.

This may seem to be mere hair-splitting—an exercise in the drawing of distinctions for its own sake. Really, however, what has just been said is related to a very important fact about language. Consider again the pictorial relationship. Discussing this in III.1, we said that Wittgenstein asserts that we establish a connexion between the elements of a picture (or, we can now add, of a propositional sign) and objects. Now, we do not (for example) connect just one occurrence of a word with an object; we do not say, 'The word which I have just uttered—and no other utterance of that word—is to be a representative of this object'. We need to be able to refer to the same object by the same word on different occasions; that is, we need to establish a connexion between an object and a word, not as a token, but as a type. Similar reasoning leads to the conclusion that the propositional sign, too, must be a type.

Exercise

Write a short account of the way in which Wittgenstein's views on meaning and sense differ from Frege's views on sense and reference.

IV.5 THE ARGUMENTS FOR THE EXISTENCE OF OBJECTS

Reading: *T* 3.23, 2.021–2.023, 2.026

We said in the last section that we are not to expect from Wittgenstein any examples of 'completely analysed' propositions; that is, we are not to expect any examples of simple signs. Nevertheless, Wittgenstein insists that simple signs must be possible (3.23). They must be possible, he says, because sense must be determinate.

He does not develop this point, but what he says is closely related to the first of the two arguments for the existence of objects offered earlier in number 2 of the *Tractatus*; indeed, as he puts it in the Notebooks (*N*, p. 63), the demand for simple things *is* the demand for determinateness of sense. Now that something has been seen of Wittgenstein's views about sense and meaning, we are in a position to discuss these arguments.

IV.5.1 The First Argument

The first argument begins (2.021) with the assertion that objects make up the substance of the world, and for that reason they cannot be composite. Wittgenstein explains this by saying (2.0211) that if the world had no substance (i.e. if there were no simple objects) then whether a proposition had sense would depend on whether another proposition was true. But if this were so, we could not construct any picture of the world, true or false—a conclusion which Wittgenstein evidently regards as false.

These are cryptic utterances, and more than one interpretation of them has been offered. The interpretation which will be followed here[1] relates Wittgenstein's argument to a problem which Russell answered by his theory of definite descriptions. (On this, compare *Realism and Logical Analysis*, A 402 Units 5–6, Section 4, especially 4.1–4.2.) To see the connexion, however, some preliminary work is necessary. It was suggested just now that the argument put forward in 2.0211 is closely connected with the assertion, made in 3.23, that simple signs must be possible because sense must be determinate. Let us begin, then, by considering determinateness of sense. We know from

---

[1] This interpretation was first stated by G. E. M. Anscombe, *An Introduction to Wittgenstein's Tractatus*, p. 49. It has recently been ably defended by R. M. White, in G. Vesey (ed.), *Understanding Wittgenstein* (Royal Institute of Philosophy Lectures, Vol. 7, 1972–3) London, 1974, pp. 17 ff.

IV.3 that the 'sense' of a proposition is a possible fact. Now, Wittgenstein would point out that there are some possible facts which would make a given proposition true, and others which would make it false. To say that a proposition has a determinate sense is to say that there is a sharp line between these, i.e. that there is no hazy intermediate region, consisting of possible facts which make the proposition neither true nor false. (Cf. *T* 4.023, 'A proposition must restrict reality to two alternatives: yes or no'; also *Notes on Logic*, 1913 (*N*, p. 97), 'The form of a proposition is like a straight line' which divides all points of a plane into right and left'.)

All this may sound quite uncontroversial—an application of the familiar logical principle that a proposition is either true or false. The problem is, how Wittgenstein gets from this to the far from uncontroversial assertion that there must be simple objects. It is at this point that the link between Wittgenstein's argument and Russell's theory of definite descriptions appears. Let us take one of Russell's examples, the proposition 'Scott is the author of *Waverley*'. From what we saw in the last paragraph, this proposition must divide possible facts into two classes, those that would make it true and those that would make it false. Now, to speak of a possible fact is to speak of a possible combination of objects (cf. III.2, IV.3). Since the proposition is about Scott, it would be natural to suppose that Scott is a constituent of each of the possible facts in question, both those that would make the proposition true and those that would make it false. For example, the possible fact that consists of Scott's passing off someone else's original manuscript of *Waverley* as his own would be one of the possible facts that would make the proposition false. In other words, it is presupposed that Scott exists, as that which the proposition is about; that is, it is presupposed that the proposition 'Scott exists' is true.

It follows from this that if the proposition 'Scott exists' were not true, then the proposition 'Scott is the author of *Waverley*' would have no meaning. But this is plainly false. Even if there had been no Sir Walter Scott, we should still have understood the proposition 'Scott is the author of *Waverley*'—just as we understand the proposition, 'Mrs. Leo Hunter was the authoress of "The Expiring Frog"', even though there never was such a person. We must recognise, then, that the possible facts that the proposition divides into two groups are not those that we first supposed. We shall have to analyse the proposition 'Scott is the author of *Waverley*' in such a way that 'Scott' no longer functions as a proper name, standing for what the proposition is about. Russell, by his theory of definite descriptions, showed how this could be done, and in a letter to Russell written in 1913, Wittgenstein endorsed this theory (*N*, p. 128).

The next step towards the conclusion that simple objects must exist is this. We have seen that we cannot say that the meaningfulness of a proposition depends on the truth of another proposition—namely, one which asserts the existence of what is talked about. We must therefore admit that propositions are, in the last analysis, about entities which are such that the question, 'Do they exist?' *does not arise*; that is, they are about entities whose existence is self-guaranteeing, entities which must exist. The final step in the argument is to show that an entity whose existence is self-guaranteeing must be a simple object. Here, the argument probably is that the arrangement of the constituents of a complex is always contingent; it only happens to be true that they are arranged in the way that they are. As Wittgenstein says later (5.634), 'Whatever we can describe at all'—sc. any fact or state of affairs: cf. 3.144, 3.221—'could be other than it is'. So the entities whose existence is self-guaranteeing, the entities which *must* exist, can only be simple objects.

Before we leave this argument, it will be worth while to note a further point which arises out of it. We have seen one of Wittgenstein's arguments for the existence of simple objects, but we have not yet considered how we are to tell whether or not something really is simple. So in II.3 we left unanswered the question, 'Is Socrates, for example, something simple?', and in III.1 and III.2 we assumed for illustrative purposes that a cat is an object. However, in the course of the argument that we have just examined we have come across a feature of objects which we can use as a test to decide whether or not something is simple.

Exercise

What do you think this feature is?

(If in need of guidance, compare Russell's account, *RLA* 100f., *LK* 242f., of why 'Romulus' is not a proper name.)

As an example, let us consider the question, 'Is a cat a simple object?' To answer this, take any proposition which would normally be regarded as being about the cat, e.g. 'The cat is on the mat', and ask this question: 'Does "The cat is on the mat" make sense, even if the cat does not exist?' Wittgenstein would say that if it does, then 'the cat' cannot be a name, i.e. the cat is not an object. To put this in another way: a cat is not an object unless the question, 'Does the cat exist?' *does not arise*. If 'the cat' is a name, then its meaning, the object of which it is the name, must exist. As this is not so, 'the cat' is not a name, in Wittgenstein's sense of the term; similarly, 'Socrates' is not a name.

IV.5.2   The Second Argument

Wittgenstein offers a second proof of the existence of objects in 2.022–2.023. He says that an imagined world, however different it may be from the real world, must have something in common with it, and this something is a form. This 'unalterable form' (cf. 2.026) is constituted by objects.

This argument seems to involve Wittgenstein's views on meaning (*Bedeutung*) rather than on sense. Let us begin with the 'form' that the real and imagined world must have in common. We know already that for Wittgenstein, form is the possibility of structure (2.033; cf. III.2). Actual and possible worlds, then, must (according to Wittgenstein) have in common the possibility of structure, and it is objects that constitute this possiblility. What seems to be meant is this. If we think of a possible world, we cannot be thinking of objects which are different from those in the real world; we must be thinking of the same objects, but as having relations which are different from those which they have in the real world. This may seem obviously false. It seems perfectly clear that we can think of a universe containing things, animate and inanimate, of a kind quite different from those to be found in the actual universe. Wittgenstein would reply that of course this is so—but such things are not what he calls objects, they are complexes of objects. In the real and in an imagined world, the objects must be the same. If they were not— if the complexes that the writer of fiction invents were not complexes made up of the objects to be found in this world—then it would not be possible to *think* of them; a description of such a possible universe would convey nothing.

It may be asked, 'Why does Wittgenstein say that what real and imagined worlds must have in common is a *form*? Why does he not just say that they have objects in common?' Perhaps the answer is this. It will be recollected that Wittgenstein says that the world is the totality of facts, not of things or objects (cf. II.2, on *T* 1.1). Now, the facts that constitute the real world differ from what would be facts in an imagined world; what the two worlds have in common are objects which are *capable of entering into* facts. This capability is what Wittgenstein calls 'form'. We may express this in another way by recalling that for Wittgenstein, form is 'the possibility of structure'. Real and imagined worlds have the possibility of structure in common, in that they share the objects whose configurations make up the structure of the real world and which, if arranged in another way, would constitute a different world.

Now that we have discussed Wittgenstein's second argument for the existence of objects, we are in a position to consider another of his arguments about objects. This is not intended to show that they exist, but is about their nature; however, it is closely related to the argument that we have just examined. We pointed out in II.5 that, according to Wittgenstein, an object can determine only a form, and not any material properties, but we deferred consideration of his argument for this conclusion. This argument is to be found in 2.0231, which follows immediately the second argument for the existence of objects. Our first task is to see what the argument is meant to establish.

Before going further, read *T*2.0231–2.0232, and re-read the first two paragraphs of II.5.

Since a material property is contrasted with a form, and since a form is a certain kind of possibility, we may take it that a material property is a property that an object actually has. We may conjecture, on the basis of 2.0232 and 2.0251, that being of a particular colour would be an example of a material property; the corresponding form would be the mere *ability* to be of this or that colour. Now, Wittgenstein does not deny that an object can have material properties, and indeed he speaks of an object as having such properties. (2.0233, 4.023. To be exact, these propositions speak of the 'external properties' of an object. But this term seems to be equivalent to 'material properties', since the two terms are used in the same way, both being contrasted with 'form' in 2.0231 and 2.0233.) An object, then, can and does have material properties; but it cannot *determine* such properties, that is (cf. II.5) there can be no object that *must have* such and such material properties.

Why, then, does Wittgenstein say this? The answer given in 2.0231 is that 'the substance of the world'—that is, as we have seen, objects—cannot determine material properties because 'it is only by means of propositions that material properties are represented—only by the configuration of objects that they are produced'. This, like its attendant proposition 2.0232, has given commentators much trouble, and any interpretation can only be tentative. Perhaps, however, the argument of 2.0231 may be expressed as follows. Suppose that someone says that objects do determine material properties; suppose, for example, he says that every object must be of a certain colour—say, blue. This would mean that an object has this property in every possible world; possible worlds would differ in respect of the ways in which blue objects are combined, but they must consist of *blue* objects. Wittgenstein would reply that this cannot be so. What must be common to the actual world and to possible worlds are, as we saw earlier, objects considered as *capable of* entering into certain configurations, not objects actually in certain configurations. But if we say that an object is blue we are uttering a proposition such that, if it is true, there exists a certain fact or state of affairs, i.e. a certain configuration of objects. So an object cannot be blue in all possible worlds. In other words, an object's being blue cannot be determined by the object, but must depend on the existence of a certain fact or state of affairs; and the same can be said of any material property whatsoever.

It remains to consider what may be meant by the remark (2.0232) that 'In a manner of speaking, objects are colourless'. We must note the cautionary phrase, 'in a manner of speaking'; Wittgenstein is not saying that objects are colourless without qualification. In the context, it seems most likely that what is meant is that no object *necessarily* has the colour that it has. We can say of objects that they must be *capable of* being of this or that colour, but we cannot say that they *must be* of this or that colour.

The student may well ask, 'Do the arguments of the last two paragraphs relate to actual philosophical controversies? Did any philosopher ever say that things must have the material properties that they do have?' The answer is that some philosophers did say this, or something close to it. The argument of 2.0231–2, as we have interpreted it, may be regarded as an attack on the idealist doctrine of internal relations, which Moore was to criticize in a famous paper 'External and Internal Relations' (*Proceedings of the Aristotelian Society*, 1919–20; reprinted in Moore's *Philosophical Studies*, London, 1922). Briefly, the doctrine of internal relations may be regarded as saying that 'every relation in which an object stands to any other is necessary to its being the object that it is' (A. J. Ayer, *Russell and Moore: The Analytical Heritage*, London, 1971, p. 156). This is of course a doctrine that concerns relations rather than qualities, such as blue; but if one regards a quality as a monadic relation (cf. II.3) then each quality also will be internal to its subject—that is, each thing *must have* the qualities that it does have.

IV.6 NAMES AND ELUCIDATIONS

Reading: *T* 3.25–3.3

We have now seen what Wittgenstein means by 'simple signs' and 'names', and why he thinks that there must exist the objects of which names are the representatives. We said in IV.5 that he does not think it a philosopher's business to give examples of names; however, he does consider the question of how we get to know their meaning.

His discussion of this question is part of a more general account of the way in which signs 'signify' (*bezeichnen*). He draws a distinction (3.26–3.261) between primitive signs, i.e. names, and signs that have a definition. Both types of sign 'signify'—i.e. both types are significant, meaningful—but they do not signify in the same way. The way in which a sign that has a definition signifies is as follows. Propositions in which such signs occur can be analysed (3.25), and this analysis involves a kind of dissection, the taking-apart of the signs that have a definition (3.26, 3.261). This process of taking-apart must terminate in signs that 'cannot be dissected any further by means of a definition' (3.26), i.e. in 'primitive signs'. Now, every sign that has a definition signifies by way of the signs by which it is defined (3.261); that is, by way of the primitive signs. The definitions serve to 'point the way', i.e. they show how we are to reach the primitive signs. In other words: a sign that has a definition is significant in so far as it can be broken down (by means of substituting for signs that have a definition the signs that define them) into primitive signs. A primitive sign, on the other hand, is significant in so far as it is a name, and as such means an object (3.203).

How, then, do we grasp the meanings of the primitive signs? We cannot dissect them, breaking them up into signs that we do understand; they are themselves the results of such a process of dissection. It might perhaps be supposed that the method employed to grasp their meaning is one of ostensive definition—that is, the name of an object is repeatedly pronounced in the presence of the object, until we realise that the sound uttered is the object's name. In his later philosophy, Wittgenstein attacked this view strongly; in the *Tractatus* it is not so much attacked as ignored. Instead, Wittgenstein states (3.263) that the meanings of primitive signs can be explained by means of what he calls 'elucidations' (*Erläuterungen*). By 'elucidations' he means propositions that contain the primitive signs, and which can be understood only if the meanings of those signs are already known. In other words, it is not a matter of getting to know the meaning of a name first, and then using it in a proposition; rather, we get to know the meanings of names *through* propositions that contain them. As Wittgenstein puts it in 3.3, 'Only in the nexus of a proposition does a name have meaning'.

It might be thought that Wittgenstein is involved in a regress here, in that he appears to be saying that if one is to know the meanings of names, one must know the meanings of the propositions in which they occur. We are then faced with the problem: how are the meanings of these propositions explained? In fact, however, there is no regress. We must recall that, for Wittgenstein, names and propositions differ in important respects. Names have meaning (*Bedeutung*), i.e. they are the representatives of objects. Propositions, on the other hand, do not have meaning of this kind. They, and only they, have 'sense' (*Sinn*: 3.3); that is, they represent, or present, possible facts. So we get to know the *meaning* of names through understanding the *sense* of propositions which contain them. This may still seem to leave a problem: for do we not need to have the sense of propositions explained to us, and how is this to be done? We shall see later, however, that Wittgenstein says that we do not always need to have the sense of propositions explained to us; indeed, this is one of his arguments for his view that the proposition is a picture (V.3).

This account of elucidation contains an idea which the logical positivists were to develop. It is often remarked that a major difference between their empiricism and what may be called the classical empiricism of Berkeley and Hume lies in the way that they proposed to test the factual significance of words. Hume, for example, would take a word in isolation, and would ask, 'Does this word stand for a perception?' For the logical positivists, on the other hand, the question was, 'Can this word be used as part of a sentence which says what can in principle be verified?'

The account of elucidation is also important in that it enables us to establish a further link between the accounts of names and of objects in the *Tractatus*. We have seen already (II.3) that Wittgenstein states that an object must be able to occur in states of affairs. This was taken to mean that an object cannot be thought of in isolation, and this is confirmed by our recent account of names. For this has shown that a name (which is something whose meaning is an object) has no meaning in isolation. It has meaning only within the context of a proposition, and it is propositions alone that have sense, i.e. represent possible facts.

IV.7   SIGNS AND SYMBOLS

Reading: *T* 3.32–3.327

Wittgenstein has not so far suggested (except perhaps in 3.1432: see IV.3) that signs can be philosophically misleading. This is first stated clearly in the context of a distinction that he draws between signs and symbols. He begins by saying (3.32) that a sign is what can be perceived of a symbol. One and the same sign can be common to two different symbols (3.321), which will therefore signify in different ways—or, as the same phrase is translated later (3.323, 3.325), have different 'modes of signification'. The nature of a 'mode of signification' is indicated by means of examples in 3.323. This states that in ordinary language it often happens:

(i) that the same word has different modes of signification, and so belongs to different symbols. For example, the word 'is' figures as the copula, as a sign of identity, and as an expression for existence. (As examples, one might cite respectively 'John is tall', 'Jacob is Israel' and 'There is a tavern in the town'.) The word 'something' provides another example of a sign that belongs to two different symbols. We speak of *something*, but also of *something's* happening. In the first case, Wittgenstein implies, we are speaking of some *thing*, and in the second of some *event*.

It also happens, Wittgenstein says:

(ii) that two words that have different modes of signification are employed in propositions in what is superficially the same way. For example, the words 'exist' and 'go' both figure as intransitive verbs, but their modes of signification are different. It is not stated in what respect they are different, but doubtless what is in mind is the point that whereas 'goes' is a predicate, 'exists' is not. (Compare the discussion of the ontological argument for the existence of God in A 303, Units 7–8; also A 402, Units 5–6, 3.3 and 3.4.) Another example is provided by the word 'identical', which figures as an adjective. Once again, Wittgenstein does not spell out his meaning, but he probably means that if 'identical' is regarded as an adjective, it may be regarded as being in the same class as a word such as 'blue'. But whereas 'blue' is a word for a quality, identity is a relation.

What, then, is a 'mode of signification'? It is not the meaning of a sign, as is made clear by the last sentence of 3.323. Here, Wittgenstein instances the proposition 'Green is green', where the first word is the proper name of a person. The first and third words, he says, not only have different meanings, but are also different symbols—that is, they have different modes of signification. They are different symbols in that the first word is a proper name and the third is an adjective.

We have seen what a mode of signification is not; what, then, is it? It will have been noticed that in listing various modes of signification, Wittgenstein makes use of a number of terms that belong to grammar—terms such as 'cupola', 'intransitive verb', 'adjective', 'proper name'. May one say, then, that the mode of signification of a word is the way in which a grammarian would classify it, or, more briefly, the grammar of the word? Certainly, to speak of the grammar of a word in this sense is not to speak of its meaning; we do not know the meaning of the word 'green', for example, if we know only that it is an adjective. But to speak simply of the grammar of a word does not go far enough. The examples given in 3.323. show that there are words between which ordinary grammar does not distinguish—e.g. certain intransitive verbs—which do not have the same mode of signification. It would be better to say, then, that the mode of signification of a word is its *logical* grammar (cf. 3.325), or perhaps even its logical form (cf. 3.327). So it could be said that although, for example, 'exist' and 'go' are both intransitive verbs, they do not share the same logical grammar, or, they are not of the same logical form.

Let us now turn to a different question. We have seen that the same sign can belong to different symbols; but how are we to tell what the symbols are? Again, if a sign happens to belong to one symbol only, how are we to tell what that symbol is? How, as Wittgenstein puts it, are we to recognize a symbol by its sign? (3.326. More literally, 'to recognise the symbol in the sign'.) Wittgenstein's reply is that we must 'observe how it is used with a sense' (3.326). This (like the original German) is not clear; we are left asking *what* is used with a sense. However, 3.327, with its reference to the 'logico-syntactical employment' of a sign, suggests that we are to observe how the *sign* is used with a sense.

It will be recalled that it is only propositions that have sense (3.3), so in effect it is being said that we must see how the word is used in the context of propositions. When we have done this, we shall see that (for example) the logical grammar of the word 'identical' is not the same as that of the word 'blue'. This also shows what is meant by saying that a sign is what can be perceived of a symbol (3.32). A symbol is not some imperceptible entity of which a sign is a perceptible manifestation; a symbol is not a special *kind* of entity. Rather, a symbol is a sign used in a certain way, used in accordance with certain rules, and the point of 3.32 is that although a sign can be perceived by the senses, the rules of its use cannot.

For the same word to have two different modes of signification, or for two words that have different modes of signification to be used in propositions in what appears to be the same way, is not regarded by Wittgenstein as a trifling matter. On the contrary, he says that it produces the most fundamental confusions, of which the whole of philosophy is full (3.324). He gives no examples, but we have already suggested one—the idea that existence is a predicate. Again, we might instance the confusion between the 'is' of existence and the 'is' of predication, partly responsible for Parmenides' view that there is no motion or change in the universe (cf. W. K. C. Guthrie, *The Greek Philosophers from Thales to Aristotle*, Methuen, London, 1950, pp. 47 ff.). To avoid such errors we must make use of a sign-language that excludes them (3.325)—i.e. does not permit them to arise. Such a language will not use the same sign for different symbols; e.g. it will have different signs for the 'is' of existence and for the 'is' of predication. Nor will this language use in a superficially similar way signs that have different modes of signification; e.g. there will not be in it the superficial resemblance that there is between 'Father Christmas does not exist' and 'Peter does not go'.

We may call such a sign-language (borrowing a term from Russell's Introduction to the *Tractatus*, p. *ix*) a 'logically perfect' language. Wittgenstein's reference to such a language poses two problems.

(i) Russell suggested (ibid.) that Wittgenstein's concern in the *Tractatus* is with the conditions for a logically perfect language. Clearly, this is one of the concerns of the book; but is it the book's exclusive, or even its main concern?

(ii) *T* 3.325, with its reference to the logical works of Frege and Russell, suggests that the construction of a logically perfect language is the task of the logician. Given that this is so, then what is the function of the philosopher?

Before we can approach these problems, we need to see more of what Wittgenstein says about pictures and propositions, and this will be our concern in the next section.

Exercise

(The following exercise will help to bring together several of the points made in the last two sub-sections.)

Distinguish between:

(i)   The meaning of a sign
(ii) The mode of signification of a sign
(iii) The definition of a sign
(iv) The elucidation of a sign.

[handwritten margin notes:]
primitive sign a name + is an object
hence its logical grammar.
Point way to primitive sign.
– prop which contain primitive signs
+ which can be understood only
if meaning of those signs
already known.

V PROPOSITIONS AS PICTURES

V.1 IN WHAT SENSE IS THE PROPOSITION A PICTURE?

Reading: *T* 4.01–4.015

In discussing Wittgenstein's account of the proposition, we found it helpful to refer to what he said earlier in the *Tractatus* about pictures, even though pictures are seldom mentioned in the passages discussed in section IV. However, in 4.01 Wittgenstein makes explicit what has hitherto been only implicit, and says that a proposition is a picture of reality. It need not, of course, be a true or accurate picture, but it is a picture or model of reality as we imagine it, as we think it to be.

A proposition, we recollect, is the sensibly perceptible expression of a thought (3.1): it is a propositional sign in its projective relation to the world (3.12), and a propositional sign consists of words, spoken or written (3.11, 3.14). It is obvious that in calling a proposition a picture, Wittgenstein is using the term 'picture' in an extended or analogical sense, just as he used an analogy when he described the proposition in terms of projection. As he himself points out (4.011), a proposition—printed on a page, for example—does not *seem* to be a picture of reality. He replies by saying that neither does musical notation seem at first sight to be a picture of music; yet it is. More than this: the groove of a gramophone record, the musical idea, the written notes, the sound waves produced by the musicians—all these are pictures. More precisely, 'all stand to one another in the same internal relation of depicting that holds between language and the world' (4.014).

All this simply reinforces the impression that Wittgenstein is using the term 'picture' in an extended sense; the question remains, in what precise sense is he using it? When we discussed Wittgenstein's account of the proposition in terms of projection (IV.2–3) we said that he employed the metaphor of projection there to bring out two points:

(i) The elements of the propositional sign are the representatives of things. (This was the force of the analogy with 'lines of projection'.)

(ii) The fact that these elements are in a certain relation represents a corresponding possible fact.

In the explanation of the use of the term 'picture' that is given in 4.011–4.0141, Wittgenstein emphasizes a different point. He speaks, it will be recalled, of the 'internal relation' of depicting. This has no connexion with the idealist doctrine of internal relations (cf. the end of IV.5.2); Wittgenstein explains later (4.122) that by an 'internal relation' he means a 'structural relation'. This is reminiscent of what we already know about logical form (cf. III.3). A picture, Wittgenstein has said, must have its logical form in common with reality (2.18), which is to say that things must be able to be related in the same way, to have the same structure, as the elements of the picture. Wittgenstein expresses this in 4.014 by saying that the gramophone record, the written notes etc. are 'all constructed according to a common logical pattern'—or, more literally, they all have in common the same logical structure (*der logische Bau*).

In sum, Wittgenstein is saying that *A* is a picture of *B* if the two have a common logical structure, an 'inner similarity' as it is called in 4.0141. The common structure in question is that the things represented by the elements of the picture can have the same structure as the elements of the picture. At this point a further question arises, a question which is not simply one of interpretation but which involves a matter of general philosophical importance. The question is, *what is it* to speak of a common structure? Wittgenstein's answer is contained in 4.0141. Here, he explains the 'common logical pattern' of which he speaks in terms of a *general rule*. He argues that to say, for example, that the musical score is a picture of the symphony is to say that there is a general rule by means of which the musician can obtain the symphony from the score. At this point, Wittgenstein makes a brief reference to what he said earlier about projection. If one figure is a projection of another, there is a general rule connecting the points of the one with the points of the other. So Wittgenstein, in 4.0141, describes the rule that the musician uses as a 'law of projection'. Once again, the notion of projection is a metaphor; what matters is the notion of a law or rule.

V.2 WHAT PROPOSITIONS ARE PICTURES?

Reading: *T* 4.01–4.012, 4.016

The passages that we have just discussed, together with some others, throw light on a problem concerning Wittgenstein's account of the proposition: namely, *what* propositions are pictures? To be specific, the problem is this. Does Wittgenstein's view that the proposition is a picture apply only to those propositions which he calls 'completely analysed'; that is (3.2, 3.202), only to propositions whose elements are names? Or could at least one element of a proposition signify a complex (cf. 3.24), and the proposition still be a picture? Take, for example, the proposition 'The cat is on the mat'. As we have seen (IV.5) the word 'cat' signifies a complex, and the same can be said of the word 'mat'; may we say, despite this, that the proposition is a picture? If we may not, then we shall find it difficult, to say the least, to provide examples of propositions which really are pictures. For we know that in a completely analysed proposition the elements of the propositional sign correspond to objects (3.2), and we also know that Wittgenstein gives no examples of objects. We may also wonder what would be the value of a theory of meaning which applies only to propositions of which no examples are given.

The question, whether in the *Tractatus* completely analysed propositions alone are pictures is a controversial one. There is some evidence which suggests that they alone are pictures. We have seen (cf. III.1) that 2.13 says that in a picture, the elements of the picture correspond to objects; they are (2.131) the representatives of objects. From this it appears that each element of a picture corresponds to just one object; there is no suggestion that an element of a picture could signify a complex, as is said of a propositional element in 3.24. On the other hand, 2.13 is not conclusive evidence that only completely analysed propositions are pictures. Wittgenstein can hardly say everything at once, and what he does say is consistent with the possibility that an element of a proposition may correspond to a group of objects. As such, it will still correspond to *objects*, though not to just *one* object.

There is, indeed, a good deal of evidence to support the view that the picture theory of the proposition is meant to apply to more than completely analysed propositions, and some of this evidence is provided by 4.011, discussed in the last section. In 4.011, which follows immediately the assertion that the proposition is a picture of reality, Wittgenstein notes that a proposition set out on the printed page does not seem to be a picture of reality. Now, had he thought that only completely analysed propositions are pictures, he could have said in effect, 'I concede that in many cases a proposition is not a picture of reality. But such propositions are those whose elements signify a complex, and I am concerned only with completely analysed propositions.' Instead, as we have seen, he concentrates on the nature of the picture, giving examples of what seem not to be pictures, yet are. One may add as further evidence the next proposition, 4.012, in which Wittgenstein says that it is obvious that a proposition of the form '$a\mathrm{R}b$' strikes us as a picture. Whether this really is obvious, we need not enquire here; what matters is that no restrictions are placed on the propositions that can serve as instances of this particular form. 'The cat is on the mat', it seems, would serve just as well as some completely analysed proposition.

This should not be taken to imply that Wittgenstein believed that every proposition is a picture. We shall later see that this is not so when we consider the propositions of logic (in VI.3). However, there seems good reason to suppose that it is not only completely analysed propositions which are regarded by Wittgenstein as pictures. In other words: it is possible to find examples of propositions to which Wittgenstein's picture theory is held to apply.

Exercises

1 Why does it matter whether or not the picture theory of meaning applies to completely analysed propositions only?

2 (Optional) In order to avoid perhaps tedious detail, we have not cited all the passages that support the view that the picture theory applies to propositions besides those that are completely analysed. Those interested in this topic may like to examine the following propositions of the *Tractatus*:

(a) 2.1 (We picture *facts* to ourselves. On facts, cf. II.2, towards the end.)

(b) 4.016 (In order to understand the essential nature of the proposition, we are to consider hieroglyphic script. Are we to suppose that the writers of this script wrote nothing but completely analysed propositions?)

(c) 4.032 (The proposition 'Ambulo' (I walk) is implied to be a picture. Is it a completely analysed proposition?)

V.3 SUPPLEMENTARY ARGUMENTS FOR THE PICTORIAL NATURE OF THE PROPOSITION

Reading: T 4.02–4.032

So far, Wittgenstein has produced two arguments for his view that the proposition is a picture; one may be called direct, and the other indirect. The direct argument simply asks us to consider the nature of a proposition; when we have done this, and when we also understand what a picture is, we shall agree that propositions are pictures. The indirect argument is that the picture theory explains how a proposition can be false, but not senseless. If I say falsely that the cat is on the mat, what I say has sense; it represents a possible fact, a way in which things *could* be arranged.

In the propositions that follow 4.016 Wittgenstein produces further arguments for the pictorial nature of the proposition. He says (4.02), 'We can see this from the fact that we understand the sense of a propositional sign without its having been explained to us'. By 'this' he seems to refer to 'the essential nature of a proposition' (4.016), that is, its pictorial character (4.021). The argument is clear. If one understands a proposition, one knows the situation that it represents (4.021). Now, one understands a proposition without having its sense explained to one. With separate words, the case is not the same; if one is to understand a word, its meaning must at some time have been explained to one (4.026). Once one knows the meaning of words, however, the propositions in which they are used do not have to be explained. Before reading *Through the Looking Glass*, I may never have met the proposition 'The White Knight is sliding down the poker', yet I understand it without the need of any explanation of its sense. It may be asked why this should show that the proposition is a picture. The answer is that I can understand the sense of the proposition in question because it, or more precisely its propositional sign, consists of elements *which have a structure*. I may not have met the elements in just this order before; nevertheless, by their order the elements represent a possible fact. The meanings of the elements have to be explained to me, but the possible fact that the elements represent by their structure does not. This seems to be what Wittgenstein means when he says in 4.03 that a proposition is *essentially* connected with the situation that it communicates, and that this connexion is precisely that it is the logical picture of the situation.

To understand a proposition, then, it is sufficient (4.024) to understand its constituents. We do not have to know whether or not the proposition is true; all that we have to know is what is the case *if* it is true—because this is what understanding a proposition *means* (ibid. Cf. 4.063, second paragraph.) Wittgenstein adds that this thesis about our understanding of propositions is supported by the way in which we translate one language into another (4.025). We do not translate each proposition of the one into a proposition of the other; we do not, as it were, translate in units of propositions. Rather, we simply translate the constituents of the propositions.

Exercise

Can you think of an argument against this view?

Someone with experience of translation may object that it is not always sufficient simply to translate the constituents of propositions. Leaving out of account idioms, which cannot be translated in a word-for-word way, it is possible for someone to look up in a dictionary each word of a sentence and still not understand what is being said. Wittgenstein would doubtless reply that such a person has not really understood each word. He would remind us of what he said in 3.3, that it is only in the context of a proposition that a name has meaning (cf. IV.6), and he would add that this is true, not only of substantives, but also of verbs, adjectives and conjunctions (cf. 4.025). To know the meaning of the word 'white', for example, is to know the use of that word in a variety of propositions. So someone who cannot translate a sentence after looking up each word in a dictionary has not grasped the use of the words, and does not really understand them. This does not mean, of course, that if someone whose native language is not English has to translate the word 'white' as it occurs in a sentence, he must previously know the sense of that sentence. This is exactly what Wittgenstein denies; all that he would assert is that the translator must know how the word is used *in sentences*.

V.4   SHOWING AND SAYING

Reading: *T* 4.022–4.024, 4.12–4.1212

Wittgenstein has argued that we can understand a proposition without explanation because, by virtue of being a picture, a proposition represents a possible fact. In 4.022 he expresses a proposition's representation of a possible fact by saying that a proposition *shows* its sense. What is shown or represented, it must be stressed, is a *possible* fact (cf. the account of the sense of a proposition in IV.3). This is why Wittgenstein says in 4.022 that a proposition, in showing its sense, shows how things stand *if* it is true.

A proposition, then, shows how things stand if it is true; it *says* that they do so stand (4.022). So, for example, the proposition 'The cat is on the mat' shows the possible fact that the cat is on the mat, and says that the cat is on the mat. None of this is really new to us after our study of the earlier parts of the *Tractatus*; what is new, and is also of the greatest importance, is the assertion made a few pages later (4.1212) that what *can* be shown *cannot* be said. Wittgenstein begins his argument for this conclusion by saying (4.12) that in order to represent reality, a proposition must have logical form in common with reality. We spoke in III.3 of the 'logical form' that a picture must have in common with reality (2.18), and we referred to this again when discussing pictures in V.1. In 2.18 Wittgenstein was speaking of 'any picture, of whatever form', and of course what he says of such pictures applies to propositions also. To say, then, that a proposition must have logical form in common with reality is to say that things must be able to be related to each other as the elements of the proposition are related to each other. To return now to 4.12. A proposition, Wittgenstein says, represents reality; indeed, propositions can represent the whole of reality. But no proposition can represent logical form—that logical form which it must have in common with reality in order to represent it. For suppose that we were to try to represent logical form; we should have to do this by means of a proposition. This means that we should have to place ourselves, with this proposition, outside logic; and this is impossible.

But although logical form cannot be represented by propositions, it is mirrored in them; they 'show' or 'display' the logical form of reality (4.121). Again, this notion is not new to us; when discussing pictures in III.2 we saw that a picture 'displays' its pictorial form, though it cannot depict it (2.172), and we saw also that logical form and pictorial form are closely related (III.3), so much so that Wittgenstein speaks of 'logico-pictorial form' (2.2). But in 2.172 Wittgenstein seemed only to say that a picture cannot depict *its own* pictorial form; in 4.121, on the other hand, he is saying that *no* proposition can represent the logical form that is common to propositions and what they represent. 'What finds its reflection in language, language cannot represent' (4.121). Such, then, is the meaning of the assertion that what can be shown, cannot be said.

There is much here that is puzzling, but before we try to unravel at any rate some of it, it is important to note that the distinction between what can be said and what can only be shown was regarded by Wittgenstein as the main point of the *Tractatus*. When Russell wrote to Wittgenstein commenting on the manuscript of the work, he received a reply (19.8.19) which stated:

> Now I'm afraid you haven't really got hold of my main contention, to which the whole business of logical propositions is only a corollary. The main point is the theory of what can be expressed (gesagt) by propositions—i.e. by language —(and, which comes to the same, what can be *thought*) and what can not be expressed by propositions, but only shown (gezeigt); which, I believe, is the cardinal problem of philosophy. (Wittgenstein, *Letters to Russell, Keynes and Moore*, Oxford, 1974, p. 71.)

This makes it clear that Russell was wrong in supposing (cf. IV.7) that Wittgenstein's main concern in the *Tractatus* was with the conditions for a logically perfect language, but it does not make clear why Wittgenstein thought the distinction between saying and showing to be of such great importance. We shall discuss this later (VI.4, VII.1–2); for the moment, however, let us consider *what* Wittgenstein is saying in 4.1212 and the propositions on which it depends.

This is to some extent obscured by his use of the word 'represent' (*darstellen*). In 4.12–4.121 there is an implied contrast between representing and showing; earlier, however, it was said (2.221) that a picture 'represents' its sense, and 'represent' seems to be used there in the sense of 'show', as when it is said that a proposition 'shows' its sense (4.022). What, then, is meant when it is said in 4.12 that propositions can 'represent' the whole of reality? Evidently that propositions can *say how* things are (cf. 3.221), can *say that* things stand in such and such relations (4.022). With this in mind, let us now try to paraphrase the argument of 4.12. Suppose a proposition to say that *a*R*b*—e.g. that the cat is on the mat. It says this by virtue of the fact that it has logical form in common with reality; that is, by virtue of the fact that the relation between its elements can be the same as that between the objects of which its elements are the representatives (in this case, the cat and the mat). Now, we cannot say *what* this logical form is. For we can only do so by means of a proposition, and this proposition must show the logical form which *it* has in common with reality. If we try to frame a proposition saying what *this* logical form is, then that proposition in turn must show the logical form which it has in common with reality—and so on, to infinity. We cannot 'station ourselves with propositions somewhere outside logic' (4.12).

Our concern here is to try to understand the *Tractatus* rather than to criticize it. There is, however, one criticism of what Wittgenstein has said which can be made to throw light on his meaning, and so it will be appropriate to discuss this. Briefly, the objection is that Wittgenstein, in what he says about showing and saying, fails to distinguish between use and mention. (On this distinction, compare Units 5–6, 4.1.) Suppose, for example, that somebody writes down a musical tune that he has heard. In so doing he is following certain rules, rules for writing down music, but he is not *stating* these rules. Nevertheless, he *could* state the rules that he has used. Doing so would not be the same as using musical notation to write down a tune; in stating the rules, musical notation would be mentioned rather than used. But the point is that the rules can be stated. Let us now put this in terms of showing and saying. Musical notation, as used to write down a tune, shows the form that it has in common with reality; in describing the rules followed in writing down the tune, one is saying what the musical notation shows—despite Wittgenstein's assertion that what can be shown, cannot be said.

Wittgenstein would probably reply that this objection misses the point. In saying that what can be shown cannot be said, he is not concerned with musical notation as such but (as the context shows) with logical form; that is, something which is common to musical notation, hieroglyphic script, sentences written in the Latin alphabet, and indeed to pictures of all kinds. Now, when one states the rules that one follows in writing a tune, one does so by means of propositions, and these propositions must themselves display logical form. One has not, after all, been able to station oneself outside logic.

Wittgenstein made essentially the same point in some notes on logic which he dictated to G. E. Moore in 1914 (*N*, p. 110). There, he distinguished between the features of a proposition which are arbitrary, and which can be said, and its logical form (or what he called its 'logical properties') which is not arbitrary and which can only be shown. He wrote:

> In any ordinary proposition, e.g., 'Moore good', this *shews* and does not say that '*Moore*' is to the left of 'good'; and *here what* is shewn can be *said* by another proposition. But this only applies to that *part* of what is shewn which is arbitrary. The *logical* properties which it shews are not arbitrary, and that it has these cannot be said in any proposition.

This can easily be applied to the example taken from musical notation. Obviously some features of such notation are arbitrary; it is, for example, arbitrary that a higher note should be expressed in the notation by a sign written higher on the same stave. Such arbitrary features, shown in the notation, can be stated in a proposition. It does not matter that this proposition will display logical form, for it is not the logical form of musical notation that it states. The situation is quite different, however, if we try to state the logical form of the notation, which is not arbitrary. Here, the fact that the propositions that we use to try to state logical form display such form themselves means that we cannot get outside logic, cannot state the logical form of the notation.

In discussing Wittgenstein's account of the proposition, we have hitherto concentrated on what he says about the proposition in general. We have mentioned the distinction between 'completely analysed' and other propositions, but we have so far said little about this. It is now time to consider in greater detail the types of proposition recognised in the *Tractatus*.

Exercise

Explain Wittgenstein's assertion (*T* 4.121) that 'what finds its reflection in language, language cannot represent'.

[handwritten notes:]

logical form is only mirrored in language
— language cannot represent it
∴ That is shown cannot be said
— we cannot state the logical form because we would have to go outside language to do so.
— impossible
— language is our world
scripture ???

VI ELEMENTARY PROPOSITIONS AND TRUTH-FUNCTIONS

VI.1 THE NATURE OF ELEMENTARY PROPOSITIONS

Reading: *T* 4.21–4.23

In 4.21, Wittgenstein introduces the technical term 'elementary proposition'. Though the term has not met us previously in the *Tractatus*, Wittgenstein is not introducing a new idea, for to speak of elementary propositions is another way of referring to what were previously called 'completely analysed' propositions. The elements of those propositions were 'simple signs' or 'names' (3.2, 3.201–3.202; cf. IV.4 above); similarly, the elementary proposition, the 'simplest kind of proposition' (4.21), consists of names (4.22). The connexion between the terms 'elementary propositions' and 'completely analysed propositions' is made explicit in 4.221, where Wittgenstein says that it is obvious that the analysis of propositions must bring us to elementary propositions. Henceforth Wittgenstein drops the term 'completely analysed' propositions, speaking of 'elementary' propositions instead, and we will do the same.

It was suggested in V.2 that Wittgenstein's picture theory of the proposition does not apply to elementary propositions alone. However, there is no doubt that he believed that elementary propositions are pictures; indeed, when in 4.0311 he compares a proposition to a picture—or more exactly to a 'tableau vivant', a 'living picture'—he does so in a way that suggests that he has elementary propositions in mind. To consider such propositions, then, will deepen our understanding of the picture theory of the proposition; we shall also find that it throws more light on Wittgenstein's reasons for his views about objects.

Between 4.21 and 4.23 Wittgenstein discusses the relations of elementary propositions to each other, and also the constituents of an elementary proposition. We shall discuss these topics in VI.1.1 and VI.1.2; then, in the light of what has been said there, we shall return in VI.1.3 to questions about the nature of objects that were raised in section II.

VI.1.1 The Relation of Elementary Propositions to Each Other

An elementary proposition, Wittgenstein says, asserts the existence of a state of affairs (4.21), and he adds (4.25) that if an elementary proposition is true, the state of affairs (sc. that whose existence it asserts) exists. (Compare the letter to Russell quoted in II.2.) We know already that an elementary proposition cannot be analysed further; in 4.211 Wittgenstein adds that there can be no elementary proposition contradicting it, or, as 6.3751 states, the logical product of two elementary propositions (that is, the joint assertion of the two; cf. II.2) cannot be a contradiction. To avoid misunderstanding, it should be noted that Wittgenstein says that elementary propositions cannot contradict *each other*; he does not say that no *proposition* can contradict an elementary proposition. Any elementary proposition—say, *p*—has a contradictory, not-*p*. But Wittgenstein would argue (see VI.2 below) that not-*p* cannot be an elementary proposition.

Wittgenstein does not explain how it is that no elementary propositions can contradict each other, but his line of argument may be as follows. Two propositions (it may be recalled from elementary logic) contradict each other if they are such that if either is true, the other is false. Now, consider the following two propositions: 'A is red all over at time t' and 'A is green all over at time t'. These are clearly contradictory; for although the propositions 'A is red all over' and 'A is green all over' are not contradictory, since they may refer to the colour of A at different times, we can say that if A is red all over at time t then it is not green all over *at the same time*, and conversely. The propositions, then, are contradictory; could they also be elementary? Wittgenstein would say that they cannot be, and his reason for this assertion may be this. Though the two propositions are contradictory, their contradictoriness is not immediately evident. For all that we can tell from the propositions as they stand, 'A is red all over at time t' and 'A is green all over at time t' might not be contradictory—just as 'A is red all over at time t' and 'A is smooth all over at time t' are not contradictory. This means that each of the two contradictory propositions must be analysable. When they are analysed, it will be found that at least one predicate (call it F) is asserted of A in the one case whose negation, not-F, is asserted of A in the other. Our examples were arbitrary, and the same line of reasoning can be applied to any pair of propositions of which one proposition contradicts the other.

To revert to our example. It may not appear obvious how one is to analyse terms such as 'red' and 'green'; one can easily understand why Locke, for example, should have called red and green 'simple ideas'. Wittgenstein would have said that the analysis can be provided by physics. In his Notebooks (N, p. 81) he says that at first sight the proposition that a point cannot be red and green at the same time does not seem to be a logical impossibility. But, he says, 'the very language of physics reduces it to a kinetic impossibility. We see that there is a difference of structure between red and green.' He is saying, in effect, that to assert that a point can be red and green at the same time is to assert that the same electro-magnetic wave can have two wave-lengths, which is an impossibility. One must be clear, however, about the nature of this reference to 'the language of physics'. Wittgenstein is not appealing to science to justify the proposition that two elementary propositions cannot contradict each other. It is *logically impossible* that they should contradict each other (cf. 6.3751), and even if physics did not provide the analysis of the contradictory propositions 'A is red all over at time t' and 'A is green all over at time t', we should still know that the propositions, because they are contradictory, must be analysable. What physics does in this case is tell us *how* they are to be analysed.

VI.1.2   The Constituents of Elementary Propositions

We noted at the beginning of VI.1 that an elementary proposition consists of names (4.22). It is 'a nexus, a concatenation, of names' (4.22); it consists of names 'in immediate combination' (4.221). This corresponds to what we have already seen about states of affairs (II.3). In a state of affairs, objects 'fit into one another like the links of a chain' (2.03), which we took to mean that objects do not require a relation to join them together. Similarly, an elementary proposition consists only of names; that is, names do not require to be joined together by any sign for a relation. It is sufficient that they are related to one another, are concatenated. We suggested earlier (IV.3) that Wittgenstein does not wish to say that a proposition such as 'The cat is on the mat' is *wrongly* expressed. However, it is now clear that, even if we suppose that 'cat' and 'mat' are names, such a proposition cannot be elementary, in that it contains words for the relation of being on.

VI.1.3 Elementary Propositions and Objects

We are now in a position to try to answer some questions about the nature of objects which have been deferred since section II. We asked in II.5.1 what examples of objects could be given; we had already asked in II.3 whether relations could be counted as examples. Let us now consider these questions. As we said in II.5.1, Wittgenstein gives no examples of objects in the *Tractatus*; however, several commentators have tried to fill this gap. It cannot be said that their labours have been wholly successful; on the other hand, they have not been entirely fruitless. It is perhaps easier to say what are not examples of objects than what are; in particular, it can be established that what Russell regarded as particulars in *The Philosophy of Logical Atomism* are not objects as the *Tractatus* understands them.

We said in II.5.1 that in *The Philosophy of Logical Atomism* Russell declared that the question of what particulars are to be found in the real world is an empirical one, of no interest to the logician. However, Russell's philosophy of logical atomism was not based exclusively on logic and the theory meaning; it was also (cf. Units 5–6, Sec. 5) influenced by epistemological considerations, namely by the answer that Russell gave to the question of what can be known. Russell makes this clear very early in his lectures, when he says that his endeavour throughout will be 'to make the views that I advocate result inevitably from absolutely undeniable data (*RLA* 32, *LK* 178 f.). Absolutely undeniable data are, for Russell, what are known, and really known; and since Russell held that the existence and nature of sense-data are known, it is not surprising that he should give sense-data as examples of particulars (*RLA* 134, *LK* 274).

Now it follows from what has been said in this section that Wittgenstein's objects, whatever they may be, do not include sense-data.

Exercise

State how, in your view, this conclusion follows.

There has been more than one type of sense-datum theory; the one with which we are concerned holds that sense-data are real things, and are indeed 'the ultimate constituents of the world' (*RLA*, loc. cit.). Similarly, the theory holds that propositions about sense-data are fundamental, in that they cannot be reduced to any other propositions. Such a proposition would be (e.g.) 'Red here and now'. Now, it is obvious that this can be contradicted by another sense-datum proposition, such as 'Green here and now'. Therefore (by what was said in VI.1.1 above) it is not an elementary proposition, from which it follows that a word for a sense-datum is not a name, and that sense-data are not objects in Wittgenstein's sense of the term.

We can now consider another question about objects. Before he propounded his philosophy of logical atomism, Russell said that in certain contexts a relation is an object (cf. VI.5 below), and we asked earlier whether Wittgenstein would count a relation as an object (II.3). What was said in VI.1.2 indicates that he would not. For if relations were objects, there would have to be names for relations, and if there were such names they could occur in elementary propositions. But it was seen in VI.1.2 that an elementary proposition consists of names that are combined 'immediately' (4.221), i.e. without being linked by a word or words for a relation.

So far, all this has been negative; can we give any positive indication of what would count as objects? Much of the controversy about the nature of objects in the *Tractatus* has revolved round the question, 'Are all objects particulars, or are at least some of them universals?' It is worth while to pause for a moment to consider why the question matters. Many philosophers have argued that universal terms, in the sense of words such as 'justice' and 'whiteness', are necessary to language. For example, in *The Problems of Philosophy* (1912) Russell said (p. 53) that no sentence could be made up without at least one word for a universal, i.e. without what we have called a 'universal term'. The problem is, how such terms have meaning. Some philosophers have thought that they have meaning in that they are the names for entities called 'universals', such as justice and whiteness. This is what is commonly called the 'realist' theory of universals, and was accepted by Russell when he wrote *The Problems of Philosophy* (pp. 56 ff.). Others have followed the 'nominalist' approach, and have said that, although there are meaningful universal *terms*, there are no such *entities* as universals; what exist are without exception particulars. There has been much debate as to whether the *Tractatus* is realist or nominalist—or, indeed, neither. It cannot be said that any agreed conclusions have been reached; however, a few tentative observations may be offered.

Though the evidence of the *Tractatus* itself does not seem to be decisive, some of Wittgenstein's later writings may give us a clue here. In some notes entitled 'Complex and Fact', written in 1931, Wittgenstein considers and rejects the view that a red circle consists of redness and circularity; he also remarks that it would be misleading to say that the fact that this circle is red is a complex of the constituents circle and redness, or that the fact that I am tired is a complex of the constituents the ego and tiredness (*Philosophische Bemerkungen*, p. 302; also in *Philosophische Grammatik*, p. 200). Wittgenstein does not say that he is here criticizing his earlier views; still, the view about the nature of a fact that is rejected is at any rate one that he *could have* held, just as he could have held a similar view about the nature of states of affairs. That this was indeed his earlier view seems to be implied by *Philosophical Investigations* I, para. 58. Here, Wittgenstein is certainly criticizing the views of the *Tractatus*, and he implies in this paragraph that 'red' is a name; as such, it must surely stand for redness. Again, when Wittgenstein criticizes the *Tractatus*' view of facts in *The Blue Book* (1933–4; Blackwell, 1958, p. 31) he gives, as examples of elements or individuals, redness, roundness and sweetness. On the basis of all this, we may conjecture that Wittgenstein's view in the *Tractatus* was that universals can occur as constituents of states of affairs.

'But', it may be objected, 'the use of these examples is quite illegitimate. For they presuppose that (e.g.) "red" is a name, as the *Tractatus* understands a name. But we have just seen that this cannot be so. Propositions such as "Red here now" or "This object is red" are not elementary, and so redness cannot be a constituent of a state of affairs, i.e. cannot be an object.' It must be admitted that 'red' is not a name; on the other hand, it is hard to believe that the *Philosophical Investigations* and *The Blue Book* entirely misrepresent Wittgenstein's earlier views. Perhaps the solution is this. Redness is not indeed an object in the *Tractatus* sense; on the other hand, it is at any rate a universal, and so Wittgenstein may have thought that it could (without being wholly misleading) serve as an example of objects as he used to conceive them.[1]

----

[1] I owe this suggestion to Dr. Stuart Brown.

VI.2   TRUTH-FUNCTIONS

Reading: *T* 4.5–5

From 4.24 onwards the *Tractatus* becomes increasingly concerned with problems about logic, and the passages that we examine in detail will be more scattered than they have been hitherto. But this does not imply that the *Tractatus* falls into two distinct and unrelated parts, one of which deals with philosophy and the other with logic. As will be seen, the philosophy and the logic are intimately connected.

Such a connexion appears when we seek an answer to the following question: 'We have seen that there are propositions that are elementary and propositions that are not elementary. How are the two classes of propositions related?' Wittgenstein's answer is short and uncompromising. He says that a proposition—*any* proposition—is 'a truth-function of elementary propositions' (*T* 5). This may seem self-contradictory. How, one may ask, can *all* propositions be truth-functions of elementary propositions? Are not elementary propositions also *propositions*? Wittgenstein would reply (*T* 5) that an elementary proposition is a truth-function of itself. Let us now see what all this means.

Wittgenstein expounds the notion of a truth-function in 4.3–4.45. He is not writing a text-book of elementary logic, and his account is not as easy to follow as it might be. For our purposes, then, it will be best to leave the text of the *Tractatus* for a moment, and give an independent exposition of the idea. (Students who have studied the Foundation Course Units, A100 L1–2, *Introduction to Logic*, will note that the definition of a truth-function given there in 4.24 differs in some respects from that given below; however, the differences are minor.) We begin by distinguishing between 'simple' and 'compound' propositions. A proposition is compound if it contains at least one other proposition as a component; a proposition is simple if it is not compound. So, for example, the proposition 'This is not white' is compound, because it contains as a component the proposition 'This is white'. Again, 'This is white or that is black' is compound, because it contains as components the propositions 'This is white', 'That is black'. On the other hand, the proposition 'This is white' is simple, because it contains no other proposition as a component. Let us call the truth or falsity of a proposition its 'truth-value'. Now, a compound proposition is a truth-function if its truth-value is determined by the truth-values of its component proposition or propositions. This definition (which corresponds to that given by Russell in *The Philosophy of Logical Atomism, RLA* 65, *LK* 209) will have to be modified if it is to fit Wittgenstein's view that *all* propositions, simple as well as compound, are truth-functions; however, it will suffice for our purposes.

The way in which the truth-value of a truth-function is determined can be represented conveniently by a table, a 'truth table'. For example, we may represent 'not-*p*' and '*p* and *q*' as follows:

p	not-*p*
T	F
F	T

pq	*p* and *q*
TT	T
FT	F
TF	F
FF	F

Under *p* in the first table, and under *p* and *q* in the second, there appear what Wittgenstein calls the 'truth-possibilities' (4.31) of the component proposition or propositions; on the right of these are written the truth-values of the compound proposition that are determined by each row of the truth-possibilities. Thus not-p is false when p is true, and true when p is false.

It is easy to see that the number of truth-functions of two simple propositions, p and q, is 16. Set out in a table (based on that given in 5.101) they are:

pq	(1)	(2)	(3)	(4)	(5)	(6)	(7)	(8)	(9)	(10)	(11)	(12)	(13)	(14)	(15)	(16)
TT	T	F	T	T	T	F	F	F	T	T	T	F	F	F	T	F
FT	T	T	F	T	T	F	T	T	F	F	T	F	F	T	F	F
TF	T	T	T	F	T	T	F	T	F	T	F	F	T	F	F	F
FF	T	T	T	T	F	T	T	F	T	F	F	T	F	F	F	F

(1) is a tautology, (2) is not both p and q, (3) is if q then p, (4) is if p then q, (5) is p or q, (6) is not q, (7) is not p, (8) is p or q but not both, (9) is if p then q, and if q then p (or, p and q are equivalent), (10) is p, (11) is q, (12) is neither p nor q, (13) is p and not q, (14) is q and not p, (15) is q and p and (16) is a contradiction.

There are two points to be made about this table, and the account of truth-functions given above. First, columns (10) and (11) show what is meant when it is said that a simple proposition is a truth-function of itself. Second, it will be noticed that we have spoken of 'simple' propositions, whereas Wittgenstein speaks of 'elementary' propositions. The point is that as far as the use of truth-tables is concerned, it does not matter to logic whether or not a proposition is elementary in Wittgenstein's sense. 'This is not white' is a truth-function of 'This is white', whether or not 'This is white' is capable of further analysis, i.e. whether or not it is elementary. This is noted by Wittgenstein, who remarks (5.31) that the truth-tables have a meaning even if the component propositions are not elementary.

Let us now return to the detailed exposition of the *Tractatus*. What has been said above about truth-functions throws more light on the nature of elementary propositions. In a letter to Wittgenstein about the *Tractatus*, Russell asked whether any elementary propositions are negative. He received the short and firm answer, 'Of course no elementary propositions are negative'. (To Russell, 19.8.19; N, p. 130.) We can see why Wittgenstein should say 'Of course'. He has said (T 5) that an elementary proposition is a truth-function of itself. But a negative proposition, e.g. not-p, is a truth-function of another proposition, namely p, and as such is not elementary.

VI.2.1   The General Form of a Proposition

We can now look at some wider issues. It will be convenient to have a name for the thesis that every proposition is a truth-function of elementary propositions; let us, then, call it the 'thesis of truth-functionality'. This thesis is closely related to what Wittgenstein says elsewhere in the *Tractatus* about what he calls 'the general form of a proposition'. In the *Philosophical Investigations* (Part I, paras. 65, 114, 134 ff.) Wittgenstein severely criticized this view. His criticisms are discussed in later Units (14–15) that deal with the *Philosophical Investigations*; our concern here is to make clear what the view criticized is.

In discussing the general form of a proposition, we have to ask first what it is that Wittgenstein is looking for; that is, what is meant by 'general form' in this context. An answer is given in T 4.5; the answer is very precise, but is also somewhat difficult. A simpler answer is given in the last two paragraphs of 5.47. This answer (which corresponds to that suggested by *Philosophical Investigations*, I, para. 65) is that what is sought is something that *all* propositions have in common with one another.

What, then, is this common feature? We will begin our account of Wittgenstein's answer by introducing the notion of an 'operation'. T 5.21 states that we can regard a proposition as the result of an 'operation' that produces it out of other propositions, which are called the 'bases' of the operation. An example will help to show what is meant. Let us suppose that p and q are both elementary propositions, and let us consider the proposition 'q and p' (column 15 of the table above). This is that proposition which is true when p is true and q is true, and is false in all other cases. We can regard this proposition as produced or generated out of (cf. 5.233) the elementary propositions p and q. p and q are the bases of the operation, and the operation itself is the construction of the appropriate truth-table. Now let us consider a certain operation which is introduced by Wittgenstein in 5.5; we will symbolize it, not as he does, but as $N(\ \ )$. The blank space within the brackets is to be regarded as keeping a place for any number of elementary propositions, and the symbol as a whole is to be regarded as expressing the following rule: 'Given any number of elementary propositions as bases, construct a truth-table which has the truth-value "truth" if and only if all the bases are false.' To illustrate. Suppose that the operation has one base, p. This would be symbolized as $N(p)$, and the result of the operation would be the truth-table

p	$N(p)$
T	F
F	T

—which is the negation of p.

If the operation has two bases, p and q, this will be symbolized as $N(p,q)$, giving as a result the truth-table

pq	$N(p,q)$
TT	F
FT	F
TF	F
FF	T

—which is the joint denial of p and q.

Similarly, if the operation has three bases, the result will be a truth-table which has the truth-value 'true' only when the three are all false; and so on, for any number of bases.

The importance of the operation $N(\ \ )$ lies in the fact that, as Wittgenstein asserts (5.5), *every* truth-function is the result of successive applications to elementary propositions of this operation. We have no space, or need, to go into the details of this; those interested are recommended to consult the lucid account given by Kenny, *Wittgenstein*, pp. 86 f. Since the thesis of truth-functionality maintains that every proposition is a truth-function of elementary propositions, it is easy to see how Wittgenstein can say (6.001) that every proposition without exception is a result of successive applications to elementary propositions of the operation $N(\ \ )$.

This, then, is what is common to all propositions—something which is stated in technical symbolism in T 6 and explained in 6.001. However, this is not the only way in which Wittgenstein explains the general form of a proposition. He also explains it in non-technical terms in 4.5, and as it is this formulation which he criticized in the *Philosophical Investigations* it is important to give it some consideration. T 4.5 states simply that the general form of a proposition is 'Es verhält sich so und so'. Pears and McGuinness translate this as 'This is how things stand', a translation which corresponds to Wittgenstein's proposed rendering of a similar phrase in 4.022 (*Letters to*

C. K. Ogden, p. 27). In the translation of the *Philosophical Investigations*, the sentence is rendered as 'This is how things are'; Wittgenstein himself preferred 'Such and such is the case' (*Letters to C. K. Ogden*, p. 30). The translation, he said (op. cit., p. 27), should not stress the word 'things'; the phrase 'es verhalt sich' is 'quite a general expression for any fact'.

Despite its apparent simplicity, this account of the general form of a proposition raises difficulties. The problem is, how it is related to the account given in T 6. One may assume that it gives a rough answer to Wittgenstein's question, on which he will refine in T 6, after he has introduced the operation which we have symbolized as $\mathcal{N}(\quad)$. But this still does not explain how precisely the two accounts are related. In seeking an answer, it is natural to look for guidance to those parts of the *Philosophical Investigations* in which Wittgenstein discusses his earlier views about the general form of a proposition. In *Philosophical Investigations* I, para. 136, Wittgenstein says that to give 'This is how things are' as the general form of a proposition is basically the same as giving the definition, 'A proposition is whatever can be true or false'. This is still a long way from what is said in T 6, but there is a connecting link—the notion of truth. As we saw, T 6 depends on the thesis of truth-functionality, and it may therefore be that when Wittgenstein said in T 4.5 that the general form of a proposition is 'This is how things stand', he meant that the common feature of all propositions is that they are truth-functions of elementary propositions.

However, this is not altogether plausible as an interpretation of 4.5, since the thesis of truth-functionality is not stated expressly until a few propositions later, in T 5. We need an interpretation of 4.5, then, which is closer to the text of the *Tractatus*; at the same time (since Wittgenstein, in his later work, can hardly be supposed to have wholly misrepresented his earlier views) it must be an interpretation which has some connexion with the one that we have derived from the *Philosophical Investigations*. With this in mind, let us look again at 4.5, and especially at its last paragraph. Wittgenstein says here that there must *be* a general form of a proposition, because 'there cannot be a proposition whose form could not have been foreseen (i.e. constructed)'. This is explained immediately afterwards, when Wittgenstein says that to speak of *all* propositions is to speak of all elementary propositions and the propositions that can be constructed out of them (4.51), or, that follow from them (4.52). (Compare the view, discussed in IV.4, that a proposition one or more of whose elements signifies a complex can be analysed into propositions each of whose elements corresponds to an object—that is, into 'completely analysed' or (VI.1) 'elementary' propositions.) It seems most probable, then, that the statement of the general form of a proposition that is given in 4.5 means that every proposition is either elementary or is constructable out of elementary propositions. As such, it is closely related to T 6, which tells us *how* all propositions can be constructed out of elementary propositions—namely, by means of the operation $\mathcal{N}(\quad)$. It is therefore also connected with (though not the same as) the thesis of truth-functionality, on which, as we saw, T 6 depends. This is how the *Philosophical Investigations* can give the impression that the general form of a proposition, as stated in T 4.5, *is* the thesis of truth-functionality. We argued earlier that this impresssion is not altogether accurate; but we have now seen that it is not far from the truth.

Exercises

To ensure that you have mastered the operation $\mathcal{N}(\quad)$, you may find it useful to do the following exercises:

Construct the truth-tables that result from the application of $\mathcal{N}(\quad)$ to (a) three and (b) four elementary propositions.

(The answers will be found in an appendix at the end of this section, on p. 73.)

VI.2.2   The thesis of truth-functionality

We have said much about the thesis of truth-functionality, but there has so far been no indication of why Wittgenstein thought that the thesis is true. The *Tractatus* does not appear to contain any explicit argument for the thesis, so it must be supposed that Wittgenstein thought its truth to be evident in the light of what is said elsewhere in the work. But in the light of what, precisely? 4.221, which states that the analysis of propositions must end with elementary propositions, seems particularly important here. In effect, 4.221 asserts that what can be said by any proposition whatever can also be said by an elementary proposition or propositions. Wittgenstein may now ask: how is this possible? Only, he may reply, if a non-elementary proposition does no more than assert or deny of a group of elementary propositions that such and such are true and such and such are false; that is, only if a non-elementary proposition is a truth-function. For example, a non-elementary proposition might assert that, of elementary propositions p, q and r, p and q are true and r is false; this is to assert p and q and not-r, which is a truth-function. Or a non-elementary proposition might deny that any one of the elementary propositions p, q and r is true; this is to assert not-p and not-q and not-r, which is also a truth-function; and so on.

The thesis of truth-functionality might seem particularly hard to refute. It says that every proposition is a truth-function of elementary propositions—and we are not given any examples of elementary propositions. However, this is not to say that the thesis is unfalsifiable. Wittgenstein does not say that no specimens of elementary propositions can be given; he says only that it is not the business of philosophy to give such examples (cf. IV.4). Nevertheless, we have seen (VI.1.3) that it is extremely difficult to say what *would* be examples of elementary propositions. Does this mean that the thesis of truth-functionality is, for practical purposes, beyond the reach of refutation? This is by no means so. There are several types of propositions which do not seem to be truth-functions of their component propositions, leaving aside the question whether these components are elementary propositions or not. If Wittgenstein cannot show that these propositions really are truth-functions of their components—if he is reduced to saying that the thesis of truth-functionality can account for these propositions in some way, he knows not what—then the thesis is seriously weakened. In fact, however, Wittgenstein does discuss the propositions that seem to tell against the thesis of truth-functionality. Of these propositions, two types are of particular philosophical importance; these are causal propositions and propositions about belief. We will discuss Wittgenstein's account of these in turn; first, however, we must consider his views about the propositions of logic, since these have a bearing on his account of causal propositions.

VI.3   THE PROPOSITIONS OF LOGIC

Reading: *T* 4.461–4.463, 6.1–6.113

In the *Tractatus*, Wittgenstein says much about the 'propositions of logic', but does not say how such propositions are to be identified. Doubtless he has in mind such works as Frege's *Begriffsschrift* and the *Principia Mathematica* of Russell and Whitehead, and if asked how we are to identify a proposition of logic he might reply, for example, 'Look at the theorems of *Principia Mathematica*; those are what I mean by "propositions of logic"'. Now, it is at once evident that these theorems are not elementary propositions; they are propositions such as 'If q, then if p then q', 'If, if p then not-q, then if q then not-p', and so on. Wittgenstein, then, must say that such propositions are truth-functions; but what truth-functions? We saw in VI.2 that one of the truth-functions is a tautology: a proposition is a tautology if it is true for all combinations of the truth-values of its component propositions. (See column (1) of the table on p. 61.) Wittgenstein asserts (6.1) that the propositions of logic are tautologies.

It will be worth while to place this remark in its historical context. *Principia Mathematica* is an axiom system, its authors' claim being that in it they deduce pure mathematics from a few axioms or 'primitive propositions' (*Principia Mathematica*, Part I, Sec. A, Preface). But why should we accept these axioms as true? The authors did not say that the axioms were true because self-evident; they did not say that the axioms were self-evident. Nor did they say (as a philosopher such as Descartes might have said) that the axioms are true because indubitable. Instead, they asserted that the reasons for accepting the axioms were *inductive*, saying (*Principia Mathematica*, Introduction, Chap. II, para. 7), 'The reason for accepting an axiom . . . is always largely inductive, namely that many propositions which are nearly indubitable can be deduced from it, and that no equally plausible way is known by which these propositions could be true if the axiom were false, and nothing which is probably false can be deduced from it'. But the authors do not explain why the propositions that can be deduced from the axioms are 'nearly indubitable', and in the preface to the first edition of the work (2nd. ed., 1927, Vol. I, p. *v*) they fall back on the vague notion of 'self-evidence'. Wittgenstein, instead of trying to justify the axioms through the theorems, puts axioms and theorems as it were on the same level; the axioms and theorems of logic are true because they are tautologies.

Whether it is correct to say that all truths of logic are tautologies is a big and still unanswered question. Our concern here is with a different and less difficult question, namely the relations between Wittgenstein's account of the propositions of logic and his philosophy in general. The connexion between this account and his views about causal propositions will appear in the next sub-section; here, we will consider its relations to his picture theory of the proposition. A tautology, Wittgenstein says, is not a picture of reality; it does not represent any possible situation (4.462). Let us consider what this means. We know that a picture represents (or 'presents') its sense (2.221), and we know that the sense of a picture is a possible fact (cf. III.2). Now, in representing a possible fact, a picture represents what may or may not be the case. For example, the proposition 'It is raining' represents a possible fact. If the proposition is true, it is raining; but it may not be raining, and in that case the proposition will be false. Now compare with this the tautology 'Either it is raining or it is not raining'. This, as Wittgenstein puts it, admits all possible situations (4.462); it is true if it is raining, but also true if it is not raining. In other words: given that I know it is raining or not raining, I know nothing about the weather (4.461). This, then, is why Wittgenstein says that a tautology does not represent any possible situation, a point which he also makes by saying that a tautology 'lacks sense' (literally, 'is senseless', *sinnlos*, 4.461). Or, as he says elsewhere, a tautology says nothing (6.11; cf. 4.461).

It is not only tautologies, says Wittgenstein, that say nothing; contradictions, too, say nothing, lack sense (4.461). To adapt the previous example: I know that it is false that it is both raining and not raining, but this tells me nothing about the weather. In 4.461, and again in 4.463 and 4.464, Wittgenstein draws a distinction between propositions on the one hand and tautologies and contradictions on the other. This can be confusing, in that in 4.46 he regards tautologies and contradictions as particular *kinds* of propositions, and in 6.1–6.12 he speaks of 'the propositions of logic', which as we know are tautologies. His reason for distinguishing propositions from tautologies and contradictions is easily seen. We know that his view is that a proposition, any proposition, is a picture of reality (4.01); but, as we have just seen, tautologies and contradictions are not pictures of reality (4.462), from which it follows logically that they are not propositions. Why, then, does Wittgenstein sometimes call them 'propositions' and why does he speak of 'the propositions of logic'? Perhaps his point is that propositions, as pictures of reality, are capable of truth and falsity; now, a tautology is at any rate *true*, and a contradiction is at any rate *false*, and to this extent they resemble pictures of reality.

One last point about tautologies. We have seen that a tautology is 'without sense', *sinnlos*; Wittgenstein denies, however, that it is 'nonsensical', *unsinnig* (4.4611). He means that if a sentence is to be nonsensical, it must break the rules of logical grammar (cf. IV.7, on signs and symbols). For example, to say 'Socrates is identical' is nonsensical in this sense (5.473), whereas a tautology such as 'If, if *p* then *q*, then if not-*q* then not-*p*' is not. The distinction is easily grasped, but the possibility of confusion might perhaps have been lessened if Wittgenstein had said, not that a tautology is 'without sense', but that it is 'empty', in that it is without content (cf. 6.111, first sentence).

VI.4   CAUSAL PROPOSITIONS AND THE LAWS OF SCIENCE

Reading: *T* 5.133–5.1361, 6.32–6.36, 6.37

We said (VI.2.2) that causal propositions are hard to reconcile with the thesis of truth-functionality, i.e. with the view that every proposition is a truth-function of elementary propositions. Let us consider briefly why this should be so. Take, for example, the proposition 'Event *A* is the cause of event *B*'. This is to say that *B* happened because *A* happened, and as such is clearly not an elementary proposition, since it contains the component propositions '*A* happened' and '*B* happened'. But neither does it seem to be a truth-function of its components. For suppose that both components are true; this would not entitle us to say that the proposition as a whole is true, nor again that it is false. For example, someone may say 'The First World War occurred because the Archduke Franz Ferdinand was assassinated at Sarajevo'. Both component propositions are true, yet the fact that they are true does not make the whole proposition true, nor again does it make it false. In short, the truth-value of a causal proposition does not seem to be completely determined by the truth values of its components; that is, a causal proposition does not seem to be a truth-function.

Wittgenstein's solution to the problem seems to consist in saying that there is no causal inference, no inference from cause *A* to effect *B*, or conversely. As he puts it (5.135) there is no possible way of inferring from the existence of any one situation to the existence of another, entirely different situation. The point of the exclusion of situations which are not entirely different from one another is easily seen. One can infer (e.g.) from the situation that the cat is on the mat and is purring to the situation that the cat is on the mat. Here, however, the situations are not entirely different and the inference is deductive, not causal. It will now be asked why Wittgenstein should say that it is impossible to infer from the existence of one situation to the existence of an entirely different situation. The answer is that to justify such an inference there would have to be a causal link, a 'causal nexus' between the two situations, and there is no causal nexus (5.136). Indeed, Wittgenstein goes so far as to say that superstition is nothing but belief in the causal nexus (5.1361). There is, he says later, no compulsion making one thing happen because another thing has happened (6.37); the *only* kind of necessity is logical necessity. That is, the only necessity is the necessity that attaches to the propositions of logic, and we have seen in VI.3 how these propositions are to be interpreted. They say nothing, are without content—whereas people who say that effect *B* necessarily follows cause *A* certainly believe that this assertion has content.

In the light of all this, let us look back at Wittgenstein's views about the mutual independence of states of affairs. We have seen (II.3, citing 2.062) that from the existence or non-existence of one state of affairs we cannot infer the existence or non-existence of another, but we have not yet asked why this should be said. We can now suggest the answer. Someone who believes that there is a link between states of affairs (Wittgenstein is in effect saying) may think of it as either causal or logical. But (i) there is no causal link, no causal nexus between states of affairs, because there is no causal nexus of any kind. (ii) There is no logical link between states of affairs, because the existence of states of affairs is asserted by elementary propositions (4.21), and one elementary proposition cannot be deduced from another (5.134). Wittgenstein does not produce an explicit defence of the latter assertion, but he may have thought that it follows from 4.211. This states (cf. VI.1.1) that no elementary proposition can contradict another. That is, it states that all elementary propositions are independent of one another; and if this is so, none can be inferred from any other.

One may be ready to concede the truth of (ii), but the truth of (i) is far less evident. (i) relies on the proposition that there is no causal nexus, and Wittgenstein cannot be said to have proved this. He can hardly argue that if there is a causal nexus the thesis of truth-functionality must be false: for the truth of this thesis is precisely what is at issue. However, Wittgenstein has more to say on the subject of causality. In general his view seems to be that the belief that there are causal propositions is a confused way of saying that there are natural laws. Now, he does not deny that there are natural laws; he does deny, however, that such laws are the *explanations* of natural phenomena (6.371), i.e. that they enable us to say that *B* happened *because* of *A*. What such laws do is *describe*.

The view that scientific laws are descriptive rather than explanatory had already been put forward by writers such as Karl Pearson. Pearson quoted the definition of mechanics given by the nineteenth-century physicist Kirchhoff: 'Mechanics is the science of motion; we define as its object the complete *description* in the *simplest* possible way of such motions as occur in nature' (Pearson, *The Grammar of Science*, 3rd ed., London, 1911, p. 115). But to say that laws of science are descriptions is to say little. The question is, what *kind* of description is, say, a law of physics? Is it merely a summary of what is observed, like the assertion that all ravens are black, or is it something else? Wittgenstein replies that it is something else. His answer may be roughly summarized as follows. The laws of physics are stated in mathematical form, and mathematics is a method of logic (6.2, 6.234). The propositions of logic, as we know, are tautologies, but Wittgenstein holds that the laws of physics are not. As he puts it (6.3431), 'The laws of physics, with all their logical apparatus, still speak, however indirectly, about the objects of the world'. (The Ogden and Ramsey translation is more literal: 'Through their whole logical apparatus the physical laws still speak of the objects of the world'.) The question is, how the laws of physics do this.

Wittgenstein replies by means of an extended analogy. Suppose, he says, (6.341) that there is a white surface, and on it there are irregular black spots. Whatever kind of picture these make, we can get as close as we like to the description of it by covering the surface with a sufficiently fine square mesh, and then saying whether each square is white or black. The mesh need not be square; we could have achieved the same result by using a triangular or hexagonal mesh. Indeed, the use of (say) a triangular mesh might have made the description simpler; that is, we might have described the surface more accurately with a coarser triangular mesh than with a finer square mesh. Now, 'the different nets correspond to different systems for describing the world'. A science such as Newtonian mechanics is, metaphorically speaking, such a net: it 'imposes a unified form' on the description of the world, just as the choice of one particular mesh imposes a unified form on the description of the white surface. Wittgenstein also makes his point by means of another metaphor. Mechanics, he says, determines one form of description of the world by saying that all propositions used in describing the world must be obtained in a given way from a given set of propositions—the axioms of mechanics. 'It thus supplies the bricks for building the edifice of science, and it says, "Any building that you want to erect, whatever it may be, must somehow be constructed with these bricks, and with these alone".'

Before we go further, there are two points to be made.

(i) Wittgenstein is not saying that Newtonian mechanics imposes a unified form *on the world*, somewhat as Kant, for example, believed that we impose space and time on the world of our experience. His point is that Newtonian mechanics imposes a unified form on our *description* of the world; i.e., given that we want to describe the world in terms of Newtonian mechanics, we must describe it in terms of such and such basic concepts, and must accept such and such axioms.

(ii) Just as there are various possible meshes, so Newtonian mechanics is only one among several forms of description of the world: not as accurate as Einsteinian physics, but more convenient for certain purposes. This is why Wittgenstein says, early in 6.341, that we can approximate *as closely as we wish* to the description of the surface.

We can now return to the question posed two paragraphs ago. This was, how do the laws of physics, through their logical apparatus, speak of the objects of the world? Or, more generally, how are the laws of physics related to the laws of logic? Wittgenstein claims (6.342) that his network metaphor enables him to give the answer. Consider again the picture that is formed by the irregular black dots. That we can describe such a picture with a net of a given form tells us nothing about the picture; for the same can be said of all such pictures. *Whatever* the picture may be, we can describe it more or less adequately by (say) a square mesh of *some* size. 'But what *does* characterise the picture is that it can be described *completely* by a particular net with a *particular* size of mesh.' Similarly, the fact that we can describe the world by means of Newtonian mechanics tells us nothing about the world; but 'what does tell us something about it is the precise *way* in which it is possible to describe it by these means'.

Although Wittgenstein says at the beginning of 6.342, 'And now we can see the relative position of logic and mechanics', some may feel that this has not been made wholly clear, and it may be helpful to look at the matter in the following way. There has been considerable controversy about the status of Newton's three laws of motion, the axioms of his mechanics. Some philosophers of science have regarded them as empirical generalizations, some of which can be tested by experiment; others have regarded them as tautologies, true by definition. Now, in saying that the possibility of describing the world by means of Newtonian mechanics tells us nothing about the world, Wittgenstein has made it clear that he does not regard Newton's axioms as empirical generalizations—for if they were, they *would* tell us something about the world. Does he regard them, then, as tautologies? Again, no; for the laws of physics speak about the objects of the world (6.3431), and tautologies, as we have seen, say nothing. In saying that Newtonian mechanics determines a form of description of the world (6.341), Wittgenstein is in effect saying that Newton's axioms are neither empirical generalizations nor tautologies; rather, they provide a *way of talking* about the world. As such, they are neither true nor false, and this is how it is that Newtonian mechanics tells us nothing about the world. It is only when we use this way of talking, and use it in a certain way—i.e. to state certain propositions and deny others—that we describe the world with the degree of accuracy that we require.

Let us recollect how we came to study Wittgenstein's account of natural laws. We were considering his view that there is no causal nexus, and we saw that this (for Wittgenstein) does not mean that there are no natural laws. But natural laws do not explain, they describe, and we have been seeing the way in which they describe. However, we have not yet completed our account of what Wittgenstein has to say about causality. Although he seems to say that there are strictly no causal laws, in that there is no causal nexus, he does say that there is, in a sense, a 'law of causality' (e.g. 6.32–6.321). One may be tempted to take this to mean, 'Every event has a cause', and the fact that Wittgenstein speaks in this context (6.34; see also a letter of Jan. 1914, *N*, p. 129) of 'the principle of sufficient reason', which is often interpreted as meaning this, may seem to support this. But it would be strange if Wittgenstein were to accept such a principle, and yet reject the idea of causal laws, and in fact he makes it clear that the 'law of causality' is, strictly speaking, the *form* of a law (6.32). As such, it might be expressed as 'There are laws of nature' (6.36).

Wittgenstein says that laws such as the principle of sufficient reason are about the net and not about what the net describes (6.35). This is not immediately clear, but as a first approximation to Wittgenstein's meaning we might suggest the following. The assertion that every event has a cause is best interpreted as saying, 'We seek for laws of nature, and we find them'. Such an assertion is not a law of nature, and so is not about what the net describes; rather, it is about the net, amounting to the assertion that the net is, or rather nets are, used successfully. But this is only a first approximation, and is not an accurate account of Wittgenstein's meaning. We suggested that the principle of sufficient reason could be interpreted as *saying* something. But, as Wittgenstein says (6.36), that there are laws of nature cannot be said, it makes itself manifest, i.e. shows itself. His reason for saying this is that, since the 'law of causality' is really the form of a law, then, like all forms, it can only be *shown*. (Cf. 4.121: propositions *show* the logical form of reality; and 4.126: what falls under a formal concept must be *shown*.) So it would be better to express Wittgenstein's view about the 'law of causality' by saying that we *show* that there are laws of nature by finding them.

Exercise

'The possibility of describing the world by means of Newtonian mechanics tells us nothing about the world' (6.342). Does this mean that Wittgenstein thought that Newton's laws of motion are tautologies?

VI.5 BELIEF

Reading: T 5.541–5.5422

It was said at the end of VI.2 that other propositions hard to reconcile with the thesis of truth-functionality are propositions about belief. Take, for example, the proposition 'A believes that p' (i.e. 'A believes that so-and-so is the case'). Suppose that p is true; we cannot derive from this anything about the truth or falsity of the proposition as a whole. A *might* be believing what is true, but equally he might not. Similarly, if p is false we cannot derive from this anything about the truth or falsity of 'A believes that p'. In sum, the truth or falsity of the proposition does not appear to be determined by the truth or falsity of its component. The same argument can be brought in the case of 'A has the thought p' and 'A says p', but we will concentrate on 'A believes that p'.

Wittgenstein states this objection in 5.541, and answers it by saying that it rests on a mistaken view about the nature of belief. Before we discuss his answer, however, it will be useful to clear up an ambiguity in such words as 'believe' and 'belief'. Often, when we say that we believe that p, we imply that we do not know that p. I may say, for example, that I believe that such and such is the correct interpretation of a passage from the *Tractatus*, implying that I do not know, or even claim to know, that it is. Sometimes, however (and especially in philosophy), to say that someone believes that p is consistent with his knowing that p. In this sense, 'believe' is synonymous with 'judge', and to 'judge' that p is to state seriously, or be prepared to state seriously, that p—seriously, in the sense that one means what one says, and is not (for example) reciting a sentence as a phonetic exercise. Such a sense of 'believe' is in mind when philosophers say that to know that p is (1) to believe that p, (2) to be certain of p, (3) to be right about the truth of p and (4) to have adequate grounds for one's belief. It is also the sense of 'believe' that is in mind in the *Tractatus* 5.541–5.5422, as can be seen from Wittgenstein's switch to the phrase 'makes the judgement p' in 5.5422. As will be seen later, the word 'believe' was also given the same sense by Russell in passages which Wittgenstein certainly had in mind here.

Let us now return to Wittgenstein's account of belief. In 5.541, he says that it had been supposed that, when A believes that p, the proposition p stands in some kind of relation to an object A. Wittgenstein replies that this is not so; really, the proposition 'A believes that p' is of the form '"p" says p' (5.542). — *No relation* This is very condensed but, in the light of what we have seen of Wittgenstein's views about the proposition, perfectly intelligible. Let us consider two possible cases: (a) a case in which A believes that p, and also utters the proposition 'p'; (b) a case in which A believes that p, but keeps his belief to himself.

In case (a), A is using a propositional sign as a projection of a possible situation; that is, he is using certain words to picture a possible situation. A propositional sign, we recall, is a fact (3.14), consisting of words which stand in certain relations to each other, and this fact represents that things are related to one another in the same way (2.15). This is how Wittgenstein can say (5.542) that '"p" says p' involves 'the correlation of facts by means of the correlation of their objects'. In case (b), in which A believes that p, but does not communicate his belief, the analysis is fundamentally the same, the difference being that in this case the constituents of the thought will not be spoken or written words, but certain of the 'psychical constituents' of which Wittgenstein spoke in a letter quoted earlier (IV.1). If it is asked how (a) and (b) help to save the thesis of truth-functionality, the reply would be that the proposition 'A is using certain words (or psychical constituents) to picture a possible fact' *can* be regarded as a truth-function of propositions about what A is and does.

In 5.541 Wittgenstein contrasts his theory of belief with theories put forward by Russell and Moore, and in 5.5422 says why he finds Russell's theory inadequate. The reference to Moore is doubtless to his essay 'The Nature of Judgement', published in 1899. This will not be discussed here; those interested in the work are referred to Gilbert Ryle's essay 'G. E. Moore's "The Nature of Judgement"', in *G. E. Moore—Essays in Retrospect*, ed. by Alice Ambrose and M. Lazerowitz, London, Allen and Unwin, 1970, pp. 89–101. The reference to Russell is to various articles and chapters dating from about 1910, and it will be worth while to pursue these here for the light they throw on Wittgenstein's views. One point must be stressed at the outset. When Wittgenstein criticizes 'Russell's theory' (5.5422) he is *not* criticizing the theory of judgement put forward in Part IV of *The Philosophy of Logical Atomism*; that theory is one that replaces the views that Wittgenstein had criticized (*RLA* 81 ff., *LK* 224 ff.). These views were first published in the essay 'On the Nature of Truth and Falsehood', which appeared in *Philosophical Essays* (1910); similar views were stated in Chap. 12 of *The Problems of Philosophy* (1912). There are also references to the theory in 'Knowledge by Acquaintance and Knowledge by Description' (1910–11; in *Mysticism and Logic*, 1963 ed., pp. 159 ff.), and in the Introduction to *Principia Mathematica*, (1st ed., 1913), Vol. I, Chap. II, sec. 3 (2nd ed., 1927, p. 43).

This theory of judgement is best approached by considering what Russell was trying to achieve. He wanted a theory which: (a) did not introduce a type of entity, called an 'idea', between the judging mind and the external world (*Mysticism and Logic*, p. 160). (b) Did not presuppose the existence of a class of entities called 'propositions'. (This is stated as a desideratum in Russell's revised theory, *RLA* p. 79, *LK* 223, but it is also true of his earlier theory.) Finally, Russell sought a theory which (c) allowed for the possibility of false judgements.

Broadly, Russell's solution of these problems was that judging or believing is a relation between a mind, which he calls the 'subject', and objects, which are other than the mind (*The Problems of Philosophy*, pp. 72 f.). This meets requirements (a) and (b), but it is not clear from this how requirement (c) is met. To see how (c) is met, it is necessary to consider more closely the relation of judging. If there are to be false judgements, Russell argued, the relation of judging must be 'multiple', i. e. must have more than two terms (*Philosophical Essays*, 1966 ed., p. 155). Take, for example, Othello's (false) belief that Desdemona loved Cassio. It cannot be said here that the relation of judging has two terms only—Othello (the subject) and Desdemona's love for Cassio. Desdemona did not love Cassio, so there was no such object as her love for Cassio. To account for false judgements, then, the relation of judging must be regarded as a multiple relation. In the example just cited, Russell says that the relation of judging knits together into one complex whole both the subject (Othello), and the objects Desdemona, loving and Cassio (*The Problems of Philosophy*, pp. 73 f.). If the belief had been true, there would have existed a complex unity, consisting of the objects of the belief, in the same order as they had in the belief, but 'with the relation which was one of the objects' (sc. love) 'occurring now as the cement that binds together the other objects of the belief' (op. cit., p. 74). But the belief was false; so although the objects existed, they did not form the requisite complex unity.

Russell later said that this theory is false because of what it says about relations. 'Loves' cannot be put on a level with 'Othello' and 'Desdemona'; loving is not an object, even in the context of a judgement (*RLA* 83, *LK* 226). Wittgenstein's objection (5.5422) was that the theory made it possible for a judgement to be a piece of nonsense, such as 'This table penholders the book' (*Notes on Logic*, $\mathcal{N}$, p. 96). It is clear how Wittgenstein's picture theory escapes this objection. Roughly, a proposition represents a possible fact; we 'use the perceptible sign of a proposition . . . as a projection of a possible situation' (3.11; cf. 2.201). As this is so, no proposition can be nonsensical.

Exercise

Before going further, write a short essay comparing Russell's theory of judgement with Wittgenstein's picture theory of the proposition.

It will be worth while to dwell a little longer on Russell's theory of the judgement and its relation to the theory about the nature of the proposition put forward in the *Tractatus*. A closer examination of the two reveals some points of similarity, and helps to remove some of the strangeness that may seem to attach to Wittgenstein's theory.

We said just now that one of Russell's concerns was to find a theory of judgement which would allow for false judgements; similarly, we saw earlier that in his account of pictures Wittgenstein has the problem of falsity very much in mind (cf. III.1 (end), III.2). Further, their concern with the problem of falsity has a similar origin—namely, the part played by objects in each theory. Russell and Wittgenstein want no intermediary between (in Russell's view) the judging mind and the objects of judgement, and (in Wittgenstein's view) the proposition and the fact, i.e. a combination of objects. So falsity becomes a problem. If I say that the earth is flat, I say something false, but meaningful; I *say* something, I judge. Yet there is no such object as a flat earth.

The solutions that Russell and Wittgenstein offer to this problem differ, yet they have something in common. For Russell, the judgement in question *is* a judgement, is meaningful, because although there is no object, a flat earth, there are the objects the earth and flatness. (For the view that universals such as flatness are objects, see *Mysticism and Logic*, 1963 ed., pp. 154–5.) What does not exist is a complex unity of the earth and flatness, and so the judgement is false. As to Wittgenstein, we have seen (IV.5) that he argues that we can think of possible worlds because there are objects. Let us consider such a possible world, a world in which the earth is flat. The proposition 'The earth is flat' is meaningful because it can in principle be analysed into propositions the elements of which are names for objects; it is false, not because of the non-existence of an object or objects, but because the relevant objects do not form the required fact or facts, i.e. are not combined in the required way.

We can take further the comparison between the two theories. We have seen that Wittgenstein's picture theory is closely connected with a view about the analysis of propositions; similarly, Russell's theory of the judgement requires certain propositions to be analysable. We said just now that, for Russell, the relation of judging links together a subject and two or more objects. Now, suppose that I judge that Julius Caesar was assassinated; what are the objects of my judgement in this case? Julius Caesar no longer exists, and we have seen that Russell is not prepared to say that one of the objects is the *idea* of Caesar. Briefly, Russell's solution is that we have to substitute for 'Julius Caesar' some *description* of Julius Caesar, e.g. 'the man whose name was "Julius Caesar"'. So the judgement becomes (*Mysticism and Logic*, p. 161), 'One and only one man was called "Julius Caesar", and that one was assassinated'. This is close to saying, as Wittgenstein would have said, that the words 'Julius Caesar' in this proposition do not function as a proper name, and that it is the task of analysis to show what their real function is.

But it would be wrong to suggest that Wittgenstein is merely offering a modified version of Russell's theory; the differences—in particular, Wittgenstein's stress on propositions, i.e. on *words* used in a certain way—are too great for that. Our aim has simply been, by sketching the historical background of the picture theory of meaning, to make the theory itself more comprehensible.

APPENDIX

Answers to exercises, VI.2.1

(a)

p	q	r	$N(p, q, r)$
T	T	T	F
F	T	T	F
T	F	T	F
F	F	T	F
T	T	F	F
F	T	F	F
T	F	F	F
F	F	F	T

(b)

p	q	r	s	$N(p, q, r, s)$
T	T	T	T	F
F	T	T	T	F
T	F	T	T	F
F	F	T	T	F
T	T	F	T	F
F	T	F	T	F
T	F	F	T	F
F	F	F	T	F
T	T	T	F	F
F	T	T	F	F
T	F	T	F	F
F	F	T	F	F
T	T	F	F	F
F	T	F	F	F
T	F	F	F	F
F	F	F	F	T

VII THE LIMITS OF LANGUAGE

VII.1 THE SELF AND THE WORLD

Reading: *T* 5.6–5.641

In discussing Wittgenstein's view that the proposition is a picture we considered his distinction between showing and saying. So far we have seen two uses of this distinction: one in the discussion of logical types (II.5) and the other in the discussion of the principle of causality (VI.4). It cannot yet be said that it has been made clear why Wittgenstein regarded this distinction as of fundamental importance; however, this sub-section and the one that follows will throw more light on this.

The distinction plays an important part in Wittgenstein's discussion of solipsism, which begins in *T* 5.6. The topic of solipsism may at first sight seem remote from the doctrines of the *Tractatus* that we have considered so far. However, Wittgenstein finds a connexion between solipsism, taken as meaning that I am my world, and a thesis about the limits of language—namely, that the limits of my language are the limits of my world. The theme of the limits of language is a very important one in the *Tractatus*; indeed, in his Preface (p. 3) Wittgenstein said that the aim of the book was 'to draw a limit to thought, or rather—not to thought, but to the expression of thoughts'. In discussing the Preface (I.4 above) we remarked on its Kantian flavour, and the thesis of the limits of language is also reminiscent of Kant. (Cf. *Kant's Copernican Revolution: Moral Philosophy*, A 202, Units 17–18, 4.7.3.) We shall find as we go further in this sub-section that there are closer links between the *Tractatus* and parts of the *Critique of Pure Reason*.

But let us leave generalities and consider the relevant texts in detail. In 5.62 Wittgenstein asks how much truth there is in solipsism. His answer is not a simple one. He does not say that solipsism is true, or again that it is false. Instead, he says that what the solipsist *means* is quite correct; the world is *my* world. But, he continues, this cannot be *said*; rather, it 'makes itself manifest' (literally 'shows itself', 5.62). To see what this means, we have to take it in its context. The conclusion that the world is my world is the culmination of an argument which begins (5.6) with the assertion that the limits of my language mean the limits of my world. The connexion with solipsism, or rather with what the solipsist means, is that the world's being my world is manifest in the fact that 'the limits of *language* (of that language which alone I understand) mean the limits of *my* world' (5.62).

All this is far from clear. In trying to explain it, let us begin by asking what is meant by 'my language'. 5.62 helps to answer this. The third sentence of this proposition indicates that for 'my language' we can substitute 'that language which alone I understand'. The translation of this passage has been the subject of controversy, in that it has been thought that Wittgenstein is referring to a language which *only I* understand, i.e. a private language. However, evidence has become available which shows that the phrase means, not the language which *I alone* understand, but the *only* language which I understand. (Cf. C. Lewy, 'A note on the text of the *Tractatus*, *Mind* 76, 1967, p. 419.) Wittgenstein is of course aware of the fact that many people understand several national languages; by 'language' in 5.62 he does not mean any national language in particular, but rather (4.001) the *totality* of propositions. One may say that this language is my language, much as English is my language; it is not my private property, but all of it is as it were at my disposal, in that I can in principle utter any of the propositions that constitute language.

In what sense, then, are the limits of my language the limits of my world, and how is this connected with the world's being my world? We will begin by suggesting the broad outlines of Wittgenstein's thesis, and will then fill in the detail. Broadly, then, Wittgenstein is arguing that the world is my world, not in that it is created by me, or is me, but in that the world and language (my language) are co-extensive; he argues also that this is something that is *shown*. Let us now develop this. We saw that Wittgenstein argues that the limits of language mean, or are, the limits of my world (5.6). Commenting on this in 5.61, Wittgenstein says that logic 'pervades' (or, as Ogden and Ramsey express it, 'fills') the world; the limits of the world are also its limits. One may ask why Wittgenstein refers to *logic* here, after talking in 5.6 about *language*. The answer is that he is concerned here with what can and cannot be thought. The limits of the world, he says, are the limits of logic in that (5.61) we cannot *think* that 'the world has this in it, and this, but not that'. This may seem absurd; it seems obvious that we can think (e.g.) that whereas the world contains France and Italy it does not contain Shangri-La and Utopia. Wittgenstein would not deny this. What he is saying is quite different, being closely connected with his second argument for the existence of objects (IV.5.2). Explaining that argument, we said that Wittgenstein would agree that we can think of things that do not exist; but, we said, these 'things' are not objects as Wittgenstein understands them. Similarly, when he speaks of 'this' and 'that' here, he has in mind this or that *object*. His argument is that if I am to think that a certain object does not exist, then I must have a name for that object. But if there is a name for an object, then that object exists—otherwise the name is meaningless, is not really a name. I cannot, therefore, really think that the world does not contain such and such objects; and what I cannot think I cannot *say* either (5.61, last sentence). All that I could produce in such a case would be a string of words without sense, and this would not be to *say* anything.

It is in this sense, then, that the limits of my language are the limits of my world; it is in this sense that the world is my world—my world, in so far as I am a language-user. It is clear that I cannot *state* these limits, stationing myself as it were outside language. These limits can only show themselves, just as logical form can only show itself (cf. V.4); they show themselves (Wittgenstein would argue) in the fact that the elementary propositions to which all propositions are reduced are p_1, p_2, p_3, p_4, ... etc.

We have now seen the sense in which the world is my world, and why this cannot be said. After speaking of '*my* world', Wittgenstein goes on to discuss a related topic, the nature of the 'I'. The discussion begins with 5.631, and ends with the last proposition of number 5, 5.641. It opens with a challenging assertion: there is, Wittgenstein says, no subject that thinks or entertains ideas (5.631). We have to ask ourselves what this means, and why it is said.

In saying that there is no subject that thinks or entertains ideas, Wittgenstein is not denying our right to use the word 'I'. What he is denying is a philosophical analysis of this term, an analysis which treats the 'I' as a substance, a kind of thing. (Hence Pears and McGuinness translate the first sentence of 5.631 freely, but justifiably, as 'There is *no such thing* as the subject that thinks or entertains ideas'.) Students of A 303, Units 5–6 (*Personal Identity*) will note that Wittgenstein's conclusion here is reminiscent of

Hume's (op. cit., 2.1). However, his argument for that conclusion is significantly different. Hume had said:

> For my part, when I enter most intimately into what I call *myself*, I always stumble on some particular perception or other, of heat or cold, light or shade, love or hatred, pain or pleasure. I never can catch *myself* at any time without a perception, and never can observe any thing but the perception. (*Treatise of Human Nature*, Bk. I, Part IV, sec. 6: Selby-Bigge ed., Oxford 1888, p. 252)

Clearly, Hume's argument is empirical; the self, the 'I', is never observed. Now let us examine Wittgenstein's argument in 5.631. Suppose, he says, that I were to write a book entitled *The World as I found it*. I should have to include a report on by body, and say which parts were subordinate to my will and which were not. This, he says, would be 'a method of isolating the subject, or rather of showing that in an important sense there is no subject; for it alone could *not* be mentioned in that book'.

It is at once noticeable that, in contrast with the passage from Hume, *T* 5.631 says that the subject *could not* be mentioned in an account of 'the world as I found it'. It is not something which might be mentioned, something for which one might sensibly look, but which one does not as a matter of fact discover; it *cannot* be mentioned. Wittgenstein seems to have in mind what Ryle was later to call 'the systematic elusiveness of "I"' (G. Ryle, *The Concept of Mind*, Chap. VI, Sec. 7). His point seems to be that if I try to describe the world as I found it, the 'I' that describes cannot itself be described. I may try to describe it; but it is I who do the describing, and *this* I is not described. However far I go in describing the I that describes the I that describes . . . etc., I cannot give a complete description of the 'I'.

Let us now go further with our examination of Wittgenstein's views about the self. We saw that *T* 5.631 says that 'in an important sense' there is no subject. Does this imply that, in a sense, there *is* a subject? 5.632 gives the answer, though the full meaning of what is said is not made clear until 5.641. Wittgenstein begins 5.632 by saying that the subject does not belong to the world. This may be taken to mean that the self is not an object (cf. *T* 1.1, 2, 2.01), and it confirms our previous suggestion that Wittgenstein is denying that the self is a substance. (The connexion with substance is that objects make up the substance of the world: *T* 2.021; cf. II.4 above.) Wittgenstein continues 5.632 by saying, 'rather, it' (sc. the subject) 'is a limit of the world'. What this means is made clearer when Wittgenstein returns to the topic of solipsism in 5.64. 'Solipsism', he says, 'when its implications are followed out strictly, coincides with pure realism.' What he means is that realism holds that the world consists of objects, and not of objects plus thinking subjects; and this is what he claims to have established. He goes on by saying (5.641) that there really is a sense in which philosophy can talk about the self 'in a non-psychological way'. What brings the self into philosophy, he says, is the fact that 'the world is my world'. We have seen that this fact is connected with the limits of language and the limits of the world, so we begin to see how Wittgenstein can say that the self is a limit of the world. This is made clearer by the last paragraph of 5.641. The self with which philosophy is concerned, Wittgenstein says, is not the human being, nor the human body, nor the human soul, with which psychology deals. Rather, philosophy deals with the self as 'the limit of the world—not a part of it'.

This may still not be wholly clear, but we can throw more light on it by comparing it with earlier philosophy. Wittgenstein's account of solipsism

Schopenhauer

(and also of ethics and 'the mystical', to be discussed in the next section) has been compared with certain parts of the philosophy of Schopenhauer, whose *The World as Will and Representation* Wittgenstein had studied in his youth. These resemblances, however, will not be explored here; partly because the relations between Schopenhauer and Wittgenstein are discussed in some detail in later units of this course (Units 27–8, *The Will*), and partly because more light is thrown on these parts of the *Tractatus* by a comparison with the philosophy of Kant. This may seem strange, since there is no evidence that Wittgenstein had a first-hand knowledge of Kant when he wrote the *Tractatus*; however, he could have had an indirect knowledge of Kant's views through Schopenhauer, who regarded himself as a Kantian. The part of Kant's philosophy relevant to the account of the 'I' in the *Tractatus* is that part of the *Critique of Pure Reason* in which Kant discusses 'the paralogisms of rational psychology' (Transcendental Dialectic, Book II, Chap. 1). 'Rational psychology' is that view which regards the self as a substance; Kant, like Wittgenstein, denies that the self is one substance among others. Nevertheless, the 'I think' which must be able to accompany all my representations—the 'transcendental unity of apperception' (Transcendental Deduction, 2nd. ed., par. 16)—is regarded by Kant as of the greatest importance. But this unity is not discoverable *in* experience; rather, it is a condition *of* experience.

See Schopenhauer on Selfless consciousness

Before we leave the topic of the philosophy of the self in the *Tractatus*, it will be helpful to compare it with Wittgenstein's views about the psychology of the self. Philosophy, Wittgenstein has said, (5.641) can talk about the self 'in a non-psychological way'; what, then, would be a psychological way of talking about the self? In 5.5421 Wittgenstein speaks slightingly of 'the superficial psychology of the present day', and says that there is no such thing as the soul, as this psychology conceives it. 5.641, however, implies that there is a place for psychology, which is described as dealing with the human soul. It may be assumed that the soul with which this psychology deals is not the soul conceived as a substance; rather, it will deal with what may be called mental activities—perceiving, thinking, etc.—as these can be investigated empirically. The psychology mentioned in 5.5421 is condemned, not for being psychology, but for being infected with bad philosophy.

VII.2 ETHICS AND 'THE MYSTICAL'

Reading: *T* 6.4–6.423, 6.432–6.522

We now reach the concluding pages of the *Tractatus* (6.4 and following), in which Wittgenstein discusses ethics and (6.44, 6.45) 'das Mystische'— literally translated as 'the mystical', a translation which the author himself preferred (*Letters to C. K. Ogden*, p. 36). This discussion is far from being a mere postscript to the *Tractatus*; on the contrary, it is of major importance. The book's point, Wittgenstein wrote to his friend Ludwig von Ficker, 'is an ethical one'. (Quoted in *Letters from Ludwig Wittgenstein with a Memoir*, by Paul Engelmann, Blackwell, Oxford 1967, pp. 143–4.) In making this point, the distinction between showing and saying plays a vital part.

Wittgenstein's discussion of ethics follows 6.4, which states that all propositions are of equal value. Commenting on this, Wittgenstein says (6.41) that there is no value in the world, because everything that happens and is the case is accidental (*zufällig*; this could also be rendered as 'contingent'). From this it follows (6.42) that there can be no propositions of ethics. This is a mere outline of an argument which is itself presented in a sketchy form. Expressed more fully, it might be put in the form of the following seven propositions:

(i)   Ethics has to do with values
(ii)   No value is contingent
(iii) Everything in the world is contingent
(iv)   Therefore no value is to be found in the world
(v)   All propositions are pictures of the world
(vi)   Therefore all propositions are pictures of what is contingent
(vii) Therefore there are no propositions of ethics.

Here, (iv), (vi) and (vii) are inferences from the rest; but these other propositions—(i), (ii), (iii) and (v) need explanation and justification.

Exercise

What do you think that propositions (i), (ii), (iii) and (v) mean, and how would Wittgenstein justify them? (Note: (iii) and (v) can be answered on the basis of what has been said in earlier parts; in answering (ii), note the reference to 'ethical laws' in 6.422, and compare Kant's views about moral laws.)

Proposition (v) is familiar to us; we know that a proposition is the sensibly perceptible expression of a thought (3.1), and a thought is a logical picture of facts (3), the totality of which constitute the world (1.1). As to (iii), we have already seen (6.37: cf. VI.4) that there is no necessity other than logical necessity, and that the propositions of logic are not pictures of reality (4.462: VI.3), i.e. of the world (compare 2.19 with 2.201). Propositions (i) and (ii), however, contain new material. (i) may seem to be obvious, but it does contain an obscurity. When Wittgenstein speaks of 'ethics', does he refer to ethics as a branch of philosophy—i.e. the philosophical study of moral judgements—or does he refer to the moral judgements themselves? Since he says later (6.421) that ethics cannot be put into words, and since philosophical ethics certainly is put into words, it is likely that Wittgenstein is referring to moral judgements. The meaning of (ii)—that no value is contingent—is less clear, but a reference made later (6.422) to 'ethical laws' may throw light on what is meant. The connexion between these and values is that one's ethical values are expressed by the moral laws that one accepts; and, as Kant put it, 'If a law is to have moral force . . . it must carry with it absolute necessity' (*Fundamental Principles of the Metaphysics of Morals*, Academy ed., p. 5, trans. Abbott).

But after all this, the whole point of Wittgenstein's argument may still not be clear; one may still be left asking about the meaning of (vii), i.e. of the assertion that there cannot be propositions of ethics. This question is perhaps most easily answered if one turns to a later proposition in the *Tractatus*, *T* 6.53, which states that what *can* be said are the propositions of natural science. We shall return to this proposition in the next section; for the moment, it is sufficient to note that it indicates that by saying that there are no propositions of ethics, Wittgenstein means that the laws of ethics mentioned in 6.422 are not like the laws of natural science, or indeed like any factual assertion. This is also why he says (6.423) that it is impossible to speak about the will in so far as it is the subject of ethical attributes. In Kantian terms, we might say that the good will, in so far as it is a *good* will, is not a subject for natural science. (Cf. A 202, Units 17–18, 4.6.4, 'The two standpoints and the two worlds'.)

The assertion that there cannot be propositions of ethics is followed in 6.42 by the assertion that 'propositions can express nothing that is higher'. This may help to clarify what is meant by the assertion in 6.4 that all propositions are of equal value: namely, that none can be said to be any higher than any other. In 6.42 the term 'higher' seems to refer to moral values, but the word may have religious connexions also, since Wittgenstein says later (6.432) that *how* things are 'is a matter of complete indifference for what is higher. God does not reveal himself *in* the world.' The point that is made here is again close to one made by Kant—namely, that there is no natural theology. One cannot argue from the existence and nature of the world to the existence and nature of God.

So far, we have seen no connexion between what Wittgenstein says in these propositions and his doctrine of saying and showing; this connexion appears later, in the propositions that follow 6.5. 6.5 contains the cryptic remark '*The riddle* does not exist'. In his letters to C. K. Ogden, Wittgenstein said (op. cit., p. 37) that by 'the riddle' he meant 'the riddle par excellence'. He seems to have in mind here what we might call the cosmic riddle, or what the nineteenth-century Darwinian Ernst Haeckel called 'the riddle of the universe'. To speak of such a riddle is another way of saying that we ask such questions as 'What is the meaning of the universe? Why is there anything at all?' These questions have a distinguishing feature which marks them off from questions such as 'How did the universe form? Was there, for example, a primitive "steady state"?', the feature being that they are beyond the capacity of science to answer. In saying that 'The riddle does not exist' Wittgenstein means that such questions are not *really* questions. As he explains in the rest of 6.5, the answer to such a question cannot be put into words—that is, it cannot be said, or, it is not a matter for the natural sciences. Consequently, the question cannot be put into words either; that is, it is not a real question.

We said earlier that 6.432—'God does not reveal himself in the world'— means in effect that one cannot argue from the existence and nature of the world to the existence and nature of God. 6.5 reinforces this conclusion; it says in effect that those questions that theologians and metaphysicians pose, and which they answer by saying, 'There must be a God', are not real questions. Wittgenstein adds that the same can be said of scepticism (6.51). Scepticism, he says, is 'obviously nonsensical, when it tries to raise doubts where no questions can be asked'.

Exercise

What do you think that Wittgenstein means here?

Wittgenstein does not explain what he means by 'scepticism' here. He could be referring to philosophical scepticism, such as that of Descartes; scepticism which asks 'What do I know?' and answers that what I know is much less than I suppose. Or he could be referring to religious scepticism, that is, to doubts about the truth of religious doctrines. In the context, the latter is much the more probable, and what Wittgenstein means will be something of this sort. The philosophical theist thinks that he knows the answer to the question, 'What is the meaning of the universe?', and his answer is related to the existence and purposes of God. The sceptic thinks that he has an answer to the same question; his answer is that there is no reason to believe that the universe has any meaning. But both theist and sceptic agree that the question which they answer is meaningful, is a real question; both are wrong, and wrong for the same reason. The question they pose has and can have no scientific answer, and so is not really a question.

Wittgenstein's attack on both the philosophical theist and the sceptic has a parallel in Kant's philosophy, and it will be worth while to consider this briefly. Kant argued that natural theology is mistaken, in that a theologian of this type speaks about God in terms that resemble those of science—terms such as substance and cause—yet uses these in such a way that they do not satisfy the conditions that scientific terms must satisfy. To be more specific, his 'substance' and his 'cause' are beyond the range of possible experience. But, Kant says, the same can be said of a certain type of atheist. Such an atheist says, for example, 'There is no substance which has the features that have been ascribed to God', as if he were saying 'There is no substance of the kind which has been called "phlogiston"'. But whereas the latter is a scientific assertion, the former is not. (See *The Critique of Pure Reason*, B 781, on the 'dogmatic' opponent of religion.)

What has been said so far in this section might give the impression that Wittgenstein regarded ethics and religion as mere superstitions, to be banished to the lumber-room and replaced by the propositions of the natural sciences. But this is far from what he meant. Like Kant, he was not attacking religion, but only certain ways of trying to defend religion. (Cf. Preface to the 2nd ed. of *The Critique of Pure Reason*, B xxxii.) This is where the distinction between saying and showing comes into play. There are, Wittgenstein says, things that cannot be put into words (6.522); they are what is mystical (literally 'the mystical'), and they *make themselves manifest*, they 'show themselves'. Whether 'the mystical' includes ethics as well is not clear from this passage, but it is clear from elsewhere that ethics which, like the mystical, 'cannot be put into words' (6.421), makes itself manifest. This is indicated by 6.421, which says that ethics is 'transcendental', an assertion which Wittgenstein also makes about logic (6.13). Logic, he says there, is a mirror-image of the world, an assertion which is linked to his earlier statement (4.121) that propositions cannot represent logical form; logical form is 'mirrored' in them, i.e. they show it. In the same way, we may infer, ethics shows itself, even though it cannot be put into words.

It can now be seen what Wittgenstein meant when, in a passage from the Preface to the *Tractatus* discussed in our Introduction (I.4), he said that the value of the book lies not only in the fact that it provides definitive solutions of problems, but also in that it shows how little is achieved once these problems are solved. The problems that are solved—problems such as the nature of the proposition, the nature of logic—do not in the least affect what a man values. Such values (cf. 6.41) do not belong to the world, they cannot be put into words. But, 'There are, indeed, things that cannot be put into words' (6.522), things that can only be shown.

VII.3   THE NATURE OF PHILOSOPHY

Reading: *T* 4.003–4.0031, 4.111–4.115, 6.53–7

At the end of our discussion of signs and symbols (IV.7) we raised, but did not answer the question, 'How did Wittgenstein, in the *Tractatus*, view the nature of philosophy?' The answer to this is to be derived from three separate parts of the work (4.003–4.0031, 4.111–4.115, 6.53–7), the last of which forms the conclusion of the *Tractatus*. It will be convenient to consider these in the order in which they appear in the book.

4.003–4.0031 take up the discussion of the pitfalls of ordinary language that is contained in the account of signs and symbols in 3.323–3.325 (cf. IV.7). There, Wittgenstein pointed to the fact that, in ordinary language, the same sign is used for different symbols, and signs that have different modes of signification are used in what appears to be the same way; and he said that in this way 'the most fundamental confusions are easily produced'. He added (3.324) that the whole of philosophy is full of such confusions. 4.003 makes much the same point, saying that most of the propositions and questions of philosophers arise from the fact that we do not understand the logic of our language. In 3.324 he had spoken of philosophy as full of confusions; in 4.003 he adds that this confusion is manifested in the shape of nonsense. 'Most of the propositions and questions to be found in philosophical works are not false but nonsensical . . . The deepest problems are in fact *not* problems at all.'

Mention was made in the Introduction (I.3) of the influence exercised by the *Tractatus* on the logical positivists. Logical positivism is discussed later in this course (Units 11–13) by Dr. Stuart Brown, but we may note here, in anticipation of this account, that the passage just quoted from the *Tractatus* is close to the views of the logical positivists about much traditional philosophy, and metaphysics in particular. The logical positivists condemned metaphysics as meaningless; the metaphysician's questions were pseudo-questions and his statements were pseudo-statements. (Cf. Units 11–13, section 4, on Carnap's 'The elimination of metaphysics through logical analysis of language'.) Their reasons for saying this were to some extent the reasons given in the *Tractatus*; that is, they regarded metaphysics as resting on a failure to grasp the logic of our language. The metaphysician, Ayer wrote, 'lapses into nonsense through being deceived by grammar' (*Language, Truth and Logic*, 1936; Penguin Books, 1971, p. 60). But there is a difference. The logical positivists' main weapon against metaphysics was the principle of verification; metaphysicians were attacked because what they say cannot be verified, even in principle, by empirical means. This principle is not used in the *Tractatus*, and although some have thought that its germ is to be found in *T*4.024, there is (as Dr. Brown argues, Units 11–13) good reason to regard this view as mistaken.

All this, however, is anticipatory; let us now return to the *Tractatus*. Wittgenstein's remark that the whole of philosophy is full of confusions does not mean that he thought philosophy to be nothing but confusion. On the contrary, he thought that there are genuine tasks for philosophy. What these tasks are is first indicated in 4.0031, where Wittgenstein says that 'all philosophy' (he must mean, all philosophy that does *not* consist of meaningless statements and meaningless questions) 'is a "critique of language"'. The phrase 'critique of language' is in quotation marks in Wittgenstein's text because it is taken from Fritz Mauthner (1848–1923), a comparatively unimportant figure who was both journalist and philosopher. (See G. Weiler, 'On Fritz Mauthner's Critique of Language', *Mind* 67, 1958, pp. 80–87.) Wittgenstein is quick to add that he does not mean by the term 'critique of language' what Mauthner did—doubtless having in mind the fact that Mauthner held language to be an inadequate instrument for knowledge, a view which is inconsistent with Wittgenstein's picture theory of the proposition and the thesis of truth-functionality. What Wittgenstein means here by a critique of language is implied by his praise of Bertrand Russell, who 'performed the service of showing that the apparent logical form of a proposition need not be its real one'.

Exercise

Give an example of this 'service' performed by Russell.

Wittgenstein does not give an illustration of this; one could cite as an example Russell's theory of definite descriptions (cf. IV.5 above), but there is a simpler example which Russell himself gave to illustrate the difference between real and apparent logical form (*Our Knowledge of the External World*, 1914; 1926 ed., p. 50; cf. A 402, Units 5–6, 3.1). As expressed in ordinary language, the propositions 'All Greeks are men' and 'Socrates is a man' appear to be of the same logical form; to be more specific, both appear to be categorical propositions. In fact, however, the first is hypothetical, and should be rendered as 'For all *x*, if *x* is Greek, then *x* is a man'. Failure to make this distinction, Russell argued, led to serious errors in logic.

It appears from this that the 'critique of language' of which Wittgenstein speaks consists in showing what is the real as opposed to the apparent logical form of propositions. This may seem hard to reconcile with a remark made later (5.5563); namely, that 'All propositions of our everyday language, just as they stand, are in perfect logical order'. Perhaps this means, however, that the propositions of our everyday language are usually *significant*; they are not, as a rule, nonsensical, as philosophical propositions are often nonsensical. But this does not imply that none of them is ever a *source* of philosophical confusion.

Wittgenstein explains his concept of philosophy more fully in 4.111 and succeeding propositions. He has already spoken of philosophy as the critique of language; in 4.112 he develops this by saying that philosophy is not a body of doctrine but an activity, the aim of which is the logical clarification of thoughts, the clarification of propositions. A philosophical work, he says, consists essentially of elucidations; it clarifies and gives sharp boundaries to thoughts that, without philosophy, are cloudy and indistinct. One may raise an objection here. '*Are* the propositions of ordinary language cloudy and indistinct until the philosopher gets to work? If a non-philosopher utters a proposition, does he not—sometimes at any rate—know perfectly well what he means?' The *Tractatus* does not give an explicit answer to this. In the *Notebooks*, Wittgenstein appears to concede the force of the objection, when he says (*N*, p. 70), 'It is clear that I *know* what I *mean* by the vague sentence' (or: by the vague proposition, *Satz*). But he then continues in a way that seems to take back this concession. Suppose, he says, I tell someone, 'The watch is lying on the table', and he replies, 'If the watch were in such and such a position, would you still say that it is lying on the table?' I should then become uncertain, and 'this shows that I did not know what I meant by "lying" *in general*'. This may be linked with a remark in the *Tractatus* (4.002) to the effect that we have languages capable of expressing every sense, without having any idea how each word has meaning or what its meaning is. One might perhaps reply, 'But do we *need* to know how each word has meaning, and what its meaning is?' The answer would be that this is not the point at issue; that point was whether in ordinary language, we know what we mean, and it has now been seen why Wittgenstein would say that we do not. However, not to leave the question unanswered, it could be replied that we do need to know how each word has meaning and what its meaning is; this is because (as is implied by paras. 3 and 4 of 4.002) if we do not know this we do not grasp the logic of our language, and the fatal consequences of this for the philosopher have already been pointed out.

We have spoken of the relations between philosophy and ordinary language; what of its relation to the sciences? Wittgenstein stresses (4.111) that it is not one of the natural sciences, but is on a different level from them. We may take this to mean that philosophy does not pronounce on the truth or falsity of the propositions of the sciences; that is a matter for scientists. What it does is comment on these propositions. But how does it comment? Wittgenstein does not say that philosophy clarifies the propositions of the sciences; what he does say (4.113) is that it sets limits to the much disputed sphere of natural science, setting limits (4.114) to what can and what cannot be

thought. Here he is looking ahead to a question which he considers in
5.6–5.62, and which we discussed in VII.1: namely, how can one set limits
to thought, how can one say 'This is thinkable, this is not'? The answer
which he gives (4.114–5) is more easily understood in the light of that
discussion. He says that philosophy sets limits to what cannot be thought by
working outwards through what *can* be thought. Philosophy will 'signify
what cannot be said, by presenting clearly what can be said' (4.115).
We may compare the way (cf. VII.1) in which the limits of language show
themselves; this corresponds to philosophy's 'working outwards through
what can be thought'.

In the last three propositions of the *Tractatus* (6.53–7), following his
discussion of 'the mystical', Wittgenstein returns to the topic of the nature of
philosophy. He says (6.53) that the correct method in philosophy would
really be to say nothing except what can be said, i.e. propositions of natural
science, and then, when someone wanted to say something metaphysical, to
demonstrate to him that he had given no meaning to some of the signs in his
propositions. The reference to the meaninglessness of metaphysical utterances
does not introduce anything that is really new; we have seen already (4.003)
that for Wittgenstein, most of the propositions to be found in philosophical
works are nonsensical. What is new, however, is the reference, in the context
of a discussion of the nature of philosophy, to the propositions of natural
science. This raises two problems, one of which is of very great importance.

(a) Wittgenstein states that the propositions of the natural sciences
constitute all that can be said. (Cf. 4.11, on 'the totality of true propositions'.)
Has he, one may ask, forgotten his earlier assertion (5.5563) that the
propositions of our everyday language are in perfect logical order? The
answer to this question can only be conjectured. One solution would be to
take Wittgenstein as using the term 'natural science' in a sense so wide as to
include the propositions of our ordinary language. But it is not clear why he
should do do; further, in a passage from the Notebooks cited in VI.1.1 he
seems to imply that the language of physics—a language whose basic terms
are, for example, 'particle', 'place' and 'time'—is different from, and
superior to, the colour-words of ordinary language, in that it makes clear a
contradiction that ordinary language obscures (*N*, p. 81; cf. *T* 6.3751). It
seems more likely that the answer to the problem is that in 6.53 the term
'say' is used in a narrow sense, and means 'saying *par excellence*'. That is,
Wittgenstein may mean that the propositions of the natural sciences are
clear, its concepts sharply defined. (Hence, perhaps, there is no reference to
philosophy clarifying scientific propositions.) So to say something, *really* to
say it (as opposed to the more or less indistinct utterances of ordinary
language), is to speak the language of the natural sciences.
(b) The second problem is much more serious, and concerns the nature of
philosophy. All that can be said, we have just been told (6.53), are the
propositions of the natural sciences; but we also know that philosophy is not
a natural science. What, then, is the status of the philosophical propositions
of the *Tractatus*?

Exercise

What answers do you think were open to Wittgenstein?

Since all that can be said consists of the propositions of the natural sciences,
and since philosophy is not a science, it follows logically that the propositions
of philosophy must say nothing. There is, however, more than one way of
saying nothing. Philosophical propositions could say nothing in that they are
without sense, empty, like the propositions of logic. Alternatively, they could

say nothing in that they are nonsensical. (For the distinction between 'without sense' and 'nonsensical', cf. VI.3.)

The former alternative may appear tempting. In the *Tractatus*, Wittgenstein states what are claimed to be *necessary truths* about objects: e.g. there *must be* objects, objects *cannot be* composite. Since he has said that the only necessity is logical necessity (VI.4), it might seem that he should say that his necessary propositions about objects are empty. Yet these propositions are surely *not* tautological; to say that objects must be simple is not like saying (e.g.) that simple objects must be simple. It is not clear whether Wittgenstein reasoned in this way, but it is certain that he chose the second alternative. The propositions of the *Tractatus*, he said (6.54), are nonsensical. As philosophical propositions, they elucidate (cf. 4.112), and the way that they elucidate is that the reader who understands Wittgenstein finally recognises them *as* nonsensical. 'He must, so to speak, throw away the ladder after he has climbed up it.' He concludes (7): 'What we cannot speak about we must pass over in silence.'

The assertion that the propositions of the *Tractatus* are to be recognised as nonsensical has long seemed an outrageous paradox, a *reductio ad absurdum* of at any rate some of the docttines of the *Tractatus*. It may also involve an inconsistency. For if Wittgenstein's own propositions are nonsensical, why should he condemn other philosophers for producing nonsensical propositions and questions (4.003)? A distinction may be attempted between illuminating nonsense and confusing nonsense, but this would be a desperate remedy. Yet whatever the inadequacies of the view that the philosophical propositions of the *Tractatus* are nonsensical, Wittgenstein is calling attention to a real problem. If one agrees (as many would) that the propositions of philosophy are not propositions of natural science, and if one agrees that they are not tautologies like the propositions of logic, then what are they? It is worth noting that the problem is not peculiar to the *Tractatus*; the principle of verification presented the logical positivists with a comparable problem. The proposition that a sentence has factual meaning if and only if what it says can be verified is not tautologous, nor is it an empirical proposition. How, then, can it be meaningful?

But let us return to the *Tractatus*. Wittgenstein's problem about the nature of philosophy was generated by his views about the nature of significant discourse, i.e. of discourse which is not nonsensical. He assumes a sharp 'either . . . or'; a significant proposition is *either* a proposition of the natural sciences, *or* it belongs to logic. It is clear that if this 'either . . . or' is dissolved, if one is prepared to allow more than two types of significant discourse, then a way is open to a solution of the problem of the nature of philosophy. But this way is not opened in the *Tractatus*.

VII.4 RETROSPECT

We have now completed our examination of Wittgenstein's *Tractatus*. It will be helpful to look back on the main stages of the discussion; or rather, not just to look back on them, but in some degree to re-think them. In presenting the *Tractatus*, we have followed fairly closely the order of the book itself, beginning with the account of objects, going on to the picture theory of the proposition, then discussing elementary propositions and the thesis of truth-functionality, and finally considering what can be shown but not said. We saw, however, that Wittgenstein presents his theory of the nature of objects before he states his reasons for this theory, and it will be worth while to re-arrange the presentation of the main points of the *Tractatus* to try to bring out more clearly the logical relations between the propositions. We will attempt the task in this, the last sub-section.

The survey which follows is not meant to be a substitute for what has gone before; it is simply meant to present in a fresh order the material that is in the units that you have just studied. For this reason, references are given only to the relevant sections of these units, and not to the *Tractatus* itself. One point should perhaps be stressed. The propositions which follow are presented as one solution to a problem; it is not suggested that they are the best solution, and they are certainly not the only one.

A   Elementary and Non-elementary Propositions

1   By a 'thought' is meant a mental act which is either true or false. IV.1.
2   By a 'proposition' is meant the expression of a thought in words. IV.2.
3   Every proposition is either elementary or analysable into elementary propositions. VI.1, VI.2.
4   To say that a proposition is analysable into elementary propositions is to say that it is a truth-function of such propositions. VI.2.
5   Since an elementary proposition is a truth-function of itself, every proposition is a truth-function of elementary propositions. VI.2.
6   An elementary proposition consists of names in certain relations to each other. VI.1.2.

B   Objects, States of Affairs and Facts

7   A name stands for a simple object, which is its meaning. IV.4.
8   A name has meaning only in the context of a proposition. IV.6.
9   An object must be capable of combining with other objects. II.3.
10   A state of affairs consists of objects in relations to each other, such that they cannot be further divided into groups of objects in relations to each other; as such, it may be called 'atomic'. II.3.
11   An elementary proposition asserts the existence of a state of affairs. VI.1.1.
12   A fact is a set of states of affairs. II.2.
13   Each object determines only its 'form'; that is, the possibility of its occurring in states of affairs. II.5.
14   Objects remain the same; it is their configuration which changes. II.4.

C   Sense and Truth

15   An elementary proposition presents a possible state of affairs, which is its 'sense'. VI.1.1, III.2, IV.3.
16   A non-elementary proposition presents a possible fact, which is its 'sense'. III.2, IV.3.
17   Because a proposition must have a determinate sense, there must be simple objects. IV.5.1.
18   There must also be simple objects so that we can speak meaningfully about possible worlds. IV.5.2.
19   A proposition is true if its sense agrees, and false if its sense disagrees with reality. III.2, IV.3.

D   The Proposition as a Picture

20   A proposition is not a special kind of entity, but is a propositional sign which is used to present or 'project' a possible fact or state of affairs. IV.2.

21   The elements of a propositional sign are words. IV.2.

22   A propositional sign is not a mere collection of words but is a fact, i.e. its elements have a certain structure. IV.3.

23   A proposition says something true or false by virtue of being a picture. V.1, V.2.

24   Because a proposition is a picture, it is possible for it to communicate a new sense to us. V.3.

25   A proposition is a picture in that its elements are correlated with objects, and the fact that its elements have certain relations to each other presents the possible fact that these objects are in similar relations. III.1, IV.2, IV.3.

26   A false proposition is meaningful because, although the state of affairs or fact presented does not exist, the objects that are correlated with the elements of the proposition exist, and by their relations the elements present a possible state of affairs or fact. IV.5.2, VI.5.

27   Every proposition has logical form in common with reality. III.3, V.1, V.4.

28   To speak of logical form is to speak of a possibility of structure; i.e. it is to say that the objects with which the elements of the proposition are correlated *can be* in relations which correspond to those between the elements. III.3, V.4.

29   Logical form can only be shown; we cannot say what the logical form of a proposition is. V.4.

E   Logic and Science

30   The propositions of logic are tautologies. VI.3.

31   Strictly, they are not propositions, since they are 'without sense' and so say nothing. VI.3.

32   All necessary propositions are propositions of logic. VI.4.

33   There are therefore no necessary causal connexions between objects. VI.4.

34   The axioms of a science are neither tautologies nor empirical propositions, but establish a vocabulary by means of which nature can be described with the required accuracy. VI.4.

F   The Limits of Language

35   The ego is not an object. VII.1.

36   To say that A believes p is not to say that an object A stands in some relation to the proposition p. VI.5.

37   The limits of my language are the limits of my world; but this cannot be said, it can only be shown. VII.1.

38   What can be said are the propositions of the natural sciences. VII.3.

39   Ethics and religion do not strictly speaking contain propositions; they show themselves. VII.2.

40   Philosophy establishes the limits of the natural sciences by showing how much can be said. VII.3.

41   Philosophy is also a critique of language, in that it clarifies propositions, correcting mistakes arising out of a failure to understand the logic of our language. VII.3.

42   The whole of philosophy is full of such mistakes. IV.7.

43   Philosophy is neither a natural science nor a branch of logic. VII.3.

44   The propositions of philosophy are therefore nonsensical, though they also elucidate. VII.3.

QUESTIONS FOR DISCUSSION

1 In *The Philosophy of Logical Atomism*, Russell compares his 'particulars' to substances. To what extent can the same comparison be drawn between substances and the 'objects' of the *Tractatus*?

2 According to the *Tractatus*, what features must a picture have in common with reality, and why?

3 What is the difference between a 'proposition' and a 'propositional sign', as these terms are used in the *Tractatus*?

4 For what reasons does Wittgenstein say in the *Tractatus* that there must be simple objects?

5 What is meant in the *Tractatus* by 'the general form of a proposition'?

6 How does Wittgenstein try to reconcile propositions about belief with the thesis of truth-functionality?

7 What did Wittgenstein mean by the remark (*T* 5.62) that 'What the solipsist *means* is quite correct'?

8 Why does Wittgenstein say (*T* 6.5) that '*The riddle* does not exist'?

INDEX

THOUGHT AND REALITY: CENTRAL THEMES IN WITTGENSTEIN'S PHILOSOPHY